Artificial Intelligence

Tencent Research Institute • CAICT
Tencent AI Lab • Tencent open platform

Artificial Intelligence

A National Strategic Initiative

palgrave
macmillan

Tencent Research Institute
Tencent Research Institute
Beijing, China

Tencent AI Lab
Tencent AI Lab
Beijing, China

CAICT
Internet Law Research Center
CAICT, Beijing, China

Tencent open platform
Tencent open platform
Beijing, China

ISBN 978-981-15-6547-2 ISBN 978-981-15-6548-9 (eBook)
https://doi.org/10.1007/978-981-15-6548-9

Jointly published with China Renmin University Press

This Palgrave Macmillan imprint is published by the registered company Springer Nature Singapore Pte Ltd.
The registered company address is: 152 Beach Road, #21-01/04 Gateway East, Singapore 189721, Singapore

Preface I

Even if we could keep the machines in a subservient position, for instance by turning off the power at strategic moments, we should, as a species, feel greatly humbled.

—Alan Turing

Artificial intelligence has once again become the focus of attention in all sectors of society. It has been 60 years since the concept of artificial intelligence was first proposed. During this period, the development of artificial intelligence experienced rises and falls. In 2016, humans lost to AlphaGo at Go, the board game considered to be the last intellectual fortress. Artificial intelligence began to gradually heat up, becoming the common object of pursuit for the government, industry, research institutions, and consumer markets. With the boost of artificial intelligence strategies and capital markets in various countries, artificial intelligence companies, products, and services are constantly emerging. The third wave of artificial intelligence has arrived, which is the result of more powerful computing capabilities, more advanced algorithms, big data, the Internet of Things, and many other factors all working together. People not only continue to search for "strong artificial intelligence" that is expected to surpass humans, but have also made great progress in developing various artificial intelligence applications (called "weak artificial intelligence") that can increase productivity and economic efficiency.

On one hand, artificial intelligence is extraordinarily hot. On the other hand, there is a deep rift in the understanding of artificial intelligence between the public and professionals, technology R&D staff, and social science researchers. It is due to the existence of this gap in understanding that, many times, the artificial intelligence that people talk about is not the same concept. This often leads to unnecessary disputes and disagreements, which neither helps the development of artificial intelligence nor benefits the exploration of the true social impact of artificial intelligence.

Elon Musk and Mark Zuckerberg's debate on the threat theory of artificial intelligence represents two kinds of typical voices. On the one side is the public discourse of futuristic concerns and warnings that public intelligence and strong artificial intelligence may be out of control and threaten human survival. On the other side, the industrial community, from the perspective of function and business, continues to explore the research and development and applications of artificial intelligence. Many fields such as autonomous driving, image recognition, and intelligent robots have made great progress. At the same time, many R&D personnel believe that artificial intelligence cannot surpass humanity, and the threat theory is entertaining groundless fears.

Looking back at the history of the development of computer technology, we will find that humans' former tools, such as computers and robots, are becoming intelligent agents with a certain degree of autonomy and are beginning to replace humans in making decisions or carrying out tasks. These things, such as driving, translation and literary creation, have always been considered to be things that cannot be accomplished by machines in reality, existing only in science fiction literature.

It can be foreseen that this transfer of decision-making will become more and more common. The economic motive behind this is that people believe or hope that AI's decisions, judgments, and actions are superior to humans, or at least about the same as humans, thus liberating humans from repeated and trivial work. Take self-driving cars as an example. In traffic, 90% of traffic accidents are related to human error, and self-driving cars equipped with GPS, radar, cameras, and various sensors are given artificial eyes and ears, and their reaction speed is faster and the judgements they make are better. It is hoped that traffic accidents caused by human factors will be completely avoided.

However, at another level, it is precisely because artificial intelligence, based on its autonomy in making decisions and taking actions, is breaking away from the category of passive tools in decision-making that its judgments and actions must conform to humanity's true intentions, morals and values, and conform to legal and ethical norms. In Greek mythology, the King of Midas got his wish for a golden touch, but tragically discovered that everything he touched would become gold, including the food he ate, his daughter, and more. Will AI become a similar form of golden touch? Household robots may slaughter pets for cooking purposes. Care robots may end patients' lives for the purpose of alleviating patient suffering, and so on.

Therefore, it can be seen that the field of artificial intelligence is naturally walking between the sciences and the humanities, which requires not only contributions from mathematics, statistics, mathematical logic, computer science, and neuroscience, but also participation from philosophy, psychology, cognitive science, law, sociology, and so on. Artificial intelligence strategies or policy documents from China, the United States, the European Union, the United Nations, and other countries or international organizations all place particular emphasis on interdisciplinary research and human perspectives in the field of artificial intelligence. The term "artificial intelligence ethics" appeared more than 15 times in the *New Generation Artificial Intelligence Development Plan* issued by China. The US *National Artificial Intelligence Research and Development Strategic Plan* lists "Understand and address the ethical, legal, and societal implications of AI" as one of its main strategic directions. A legislative proposal from the European Union considers that artificial intelligence requires ethical principles and calls for the establishment of a so-called *Charter on Robotics*. The United Nations has issued a *Preliminary Draft Report on Robotics Ethics*, which states that not only do robots need to respect the ethical norms of human society, but also we need to embed specific ethical principles into robot systems. In the future, the huge importance of multi-disciplinary and multi-dimensional research and exploration of artificial intelligence will gradually become apparent.

This text *Artificial Intelligence*, written by Tencent's research institute, AI Lab, and Open Platform team in conjunction with China Academy of Information Communications Technology's Internet Law Research Center, is just such an interdisciplinary attempt. This book systematically studies the technological process, industrial trends, strategic design, legal issues, ethical issues, regulatory governance, and future imaginings related to artificial intelligence. It covers almost all the hotspots and frontier issues in the field of artificial intelligence. It is hoped that this book will promote interdisciplinary thinking, communication, and exploration in the field of artificial intelligence. Because of the limitations of our professional field and perspective, it is difficult for this book to be comprehensive, and there are inevitably mistakes or inadequacies. I invite readers to criticize and correct.

Finally, just as I quoted Alan Turing's words in the beginning, whether engaging in the research and development of artificial intelligence technology, or the development of interdisciplinary exploration and research across public policy, law, ethics, and other humanities, all must hold artificial intelligence in awe. To borrow the words of Charles Dickens, the British writer: "It was the best of times, it was the worst of times." We hope that we can all grasp this era and jointly build a bright future for artificial intelligence.

Si Xiao, Dean of Tencent Research Institute

Preface II

The increase in computing power, explosions in data, improvements in machine learning algorithms, and increased level of investment are key factors in the rapid development of a new generation of artificial intelligence. The transformation and evolution of the real economy through digitization, networking, and intelligentization have brought tremendous historic opportunities for artificial intelligence, revealing very broad prospects for development. At present, artificial intelligence products such as self-driving cars, industrial robots, smart medical devices, drones, and smart home assistants are emerging. The level of integration of artificial intelligence with various sectors of the economy and society is constantly increasing, as new models, new formats, and new technologies are building a new momentum for economic and social development, and entrepreneurial innovation is increasingly active. In the historical process of a new round of scientific and technological revolutions and industrial changes, artificial intelligence will play an increasingly important role.

The major countries in the world attach high importance to the development of artificial intelligence. The US White House has successively issued three government reports on artificial intelligence. It was the first country in the world to raise the development of artificial intelligence to the national strategic level. From the perspective of the United States, the artificial intelligence strategic plan is regarded as a new Apollo mission. The United States hopes to have the same dominance in the field of artificial intelligence as it did in the Internet age. The United Kingdom accelerated the application of artificial intelligence technology through the

Robotics and Autonomous Systems 2020 Strategy; the EU launched the world's largest civilian robot R&D program, SPARC, in 2014; the Japanese government formulated the *Japan's Robot Strategy: Vision, Strategy and Action Plan* in 2015, in order to promote the development of artificial intelligence robots. China has released the *New Generation Artificial Intelligence Development Plan* to build the first-mover advantage in artificial intelligence and accelerate the construction of an innovative country and a global technology power.

The influence of artificial intelligence is global and revolutionary. It will bring about a series of economic, social, legal, and regulatory issues, and may even subvert the existing governance system. At present, problems over conflicts and gaps between the development of artificial intelligence and relevant laws are beginning to appear, and the degree of societal attention continues to increase. Strengthening research on relevant laws, ethical and social issues, and establishing laws, regulations, and ethical and moral frameworks that guarantee the healthy development of artificial intelligence are major propositions that deserve attention.

The China Academy of Information and Communication Technology has made positive progress in the research of artificial intelligence industry, policy, law, and supervision. It has supported the *Guiding Opinions on Actively Promoting the "Internet Plus" Action Plan* and *"Internet Plus" and AI Three-Year Implementation Plan* and many other national research and drafting policies. This book is a joint research product in the field of artificial intelligence of the Internet Law Research Center of the China Academy of Information and Communication Technology and the Tencent Research Institute. This book provides a comprehensive introduction to the evolution of artificial intelligence, industry development, and the artificial intelligence policies of various countries. It analyzes legal and ethical issues, proposes governance ideas, and forecasts the development trend of artificial intelligence. I hope this book can become a window for government departments, Internet companies, research institutes, and other people from all walks of life to learn more about artificial intelligence and play an active role in advancing the development of China's artificial intelligence industry and the construction of laws and policies.

Lu Chuncong, Director of Policy and Economic Research Institute,
China Academy of Information and Communications Technology

Preface III

Consciousness isn't a journey upward, but a journey inward.
—Jonathan Nolan, *Westworld*

Artificial intelligence is not a new thing. As early as 1956 at the Dartmouth conference, the concept of artificial intelligence was formally proposed. It has been 60 years since then, and the explosion of artificial intelligence began in the past three years. The number of artificial intelligence companies born in 2015–2016 exceeds the sum of those born in the past ten years and the amount of financing is constantly breaking records. Today's humans have ushered in a truly intelligent revolution. All this is due to technological breakthroughs and mass diffusion.

When we talk about the intelligence revolution, what should we do?

One of the things Tencent's open platform is always doing is exploring the future through a vertical and horizontal "T-shaped strategy." "One vertical" represents the future advanced direction of productive forces, such as artificial intelligence, which goes deep down into the main axis of human advanced productive forces; "one horizontal" represents Tencent's open ecosystem built in the past six years, horizontally integrating resources, constantly transforming and innovating business models, in order to cultivate a fertile soil and allow productive forces to grow.

In 2017, Tencent's open platform integrates Tencent's internal AI capabilities and industry resources to realize the connection of technology and use cases, software and hardware, talent and capital, fostering a fertile soil for artificial intelligence companies, and eagerly waiting for this soil to grow towering, giant trees.

In the course of searching for artificial intelligence partners and advancing Tencent's AI accelerators, we have come across many high-quality artificial intelligence companies. Some have core artificial intelligence technologies and capabilities, and some have unique industry advantages in application scenarios. These companies are distributed in various fields such as transportation, medical care, translation, security, manufacturing, and law. The penetration of artificial intelligence at this stage and the applications that can be achieved are much richer than we think. Tencent itself is also exploring the application of artificial intelligence in various fields: serving content entrepreneurs, letting technology shine a light on culture; launching the Miying medical imaging product, so that early-stage cancer is no longer hard to discover… when artificial intelligence and industry segments are combined, a more powerful force will erupt.

Industry holds different views on artificial intelligence. Softbank's Son just feels that "sleeping is a waste of time." Tesla's Musk believes that artificial intelligence is "the biggest risk we face as a civilization." There is a line in *Westworld*, "Consciousness isn't a journey upward, but a journey inward." Artificial intelligence is created by humans, and its direction will also depend on the collective consciousness of human beings. Undoubtedly, artificial intelligence will eventually open up a new world. You can choose to wait and see, you can also choose to join in, and this book is likely to be a key to the new world.

Wang Lan, General Manager of Tencent Makerspace
Tencent Open Platform Deputy General Manager

Contents

Part I Technology: The Reality of Disruptive Technology	1
1 Gaps in Understanding of AI	3
Artificial Intelligence Rises Again	3
Gaps in Understanding About AI	4
Understanding and Recognition of Artificial Intelligence	6
Future Predictions About Artificial Intelligence	7
AI Trust and Acceptance	8
The Threat of AI	11
Legal and Research Responsibilities Relating to AI	12
Misconceptions About AI	12
Welcoming the Future	14
Bibliography	14
2 Artificial Intelligence's Past	15
Layers of AI	16
Technological Developments	23
The Third Wave of AI	30
Bibliography	30
3 Artificial Intelligence: Today and in the Future	33
Speech Processing	33
Natural Language Processing	37
Machine Learning	41

Ubiquitous Artificial Intelligence Algorithms 43
The Future of Artificial Intelligence 43

Part II Industry: The Complete Picture of the Development of AI 47

4 An Overview of the Artificial Intelligence Industry 49
The United States Is Leading the World in AI Companies 50
China and the United States Have Their Own Advantages in the Main AI Hotspots 52
US Leading Industry Giants Have First-Mover Advantage 52
What Is the Future of China's AI Industry? 53

5 Autonomous Driving 55
The Elements of Autonomous Driving 56
Levels of Autonomous Driving Technology 56
Two Roads Toward Autonomous Driving 56
The Software and Hardware Involved in Autonomous Driving 57
Trends in Autonomous Driving 60
When Will Self-driving Cars Be Road-ready? 64
Bibliography 66

6 Intelligent Robots 69
What Is a Robot? 69
The Applications of Industrial Robots Are Maturing and Steadily Growing 70
The Use of Robotics in the Service Industry Is Still in Its Infancy 72
Trends in International Robotics Development 74
Trends in the Development of China's Robotics Industry 75
Bibliography 76

7 Smart Healthcare 77
Core Smart Medical Uses 77
Examples of Smart Medicine Applications 79
Chinese Smart Healthcare Development 83

8 **AI-powered Investment Advice** 85
What Is AI-powered Investment? 85
Factors Leading to the Rise of AI-powered Investment Advice 88
The Business Model of AI-powered Investment Advice 89

9 **Smart Homes** 93
Smart Homes Are Displaying Strong Vitality on the Global Scale 93
The United States Is the Sole Leader Spearheading the Industry Development Trend 94
China's Potential Room for Expansion Is Enormous, Is the Market Window About to Arrive? 94
In Competition Over Smart Homes, Leading Companies Are Raring to Go 95
The Development Prospects of the Smart Home Industry 98
Optimizing and Improving Single Products, Expanding Application Scenarios 98
Standards Are Tending Toward Unification, the Ecology Is Gradually Maturing 98
The Issue of Smart Home Security Faces Challenges 99
The Interoperability and Interconnectedness of the Smart Home Is the Best Application Scenario for AI 100
The Trend of Smart Home Development Will Continue to Improve 100
Bibliography 101

10 **Unmanned Aerial Vehicles** 103
The Vacant State of International Unmanned Aerial Vehicles Development 103
Different Countries Have Different Development Advantages 104
Unmanned Aerial Vehicle Applications Are More Extensive 105
Spraying Drones Have Already Been Applied to Near-Mass-Scale 107
Unmanned Aerial Vehicles for International Humanitarian Relief 107
The Development of History of Unmanned Aerial Vehicles in China 108
China's Unmanned Aerial Vehicle Development Trend 109
Bibliography 111

11 Artificial Intelligence Enterprises 113
The AI Enterprise Ecosystem 113
The Classification of Chinese Artificial Intelligence Entrepreneurial Projects 115
With the Ability of Artificial Intelligence as the Foundations, Combining with Traditional Enterprises 120

Part III Strategy: A Detailed Look at National Strategies 125

12 Top-Level Plans 127
A Vast World, Full of Promise 128
World Powers "Devise Their Battle Plans 129
Bibliography 140

13 The Power of Capital 141
Funding Is the Foundational Guarantee for the Vigorous Development of Artificial Intelligence 142
Governments Increase Investment in Artificial Intelligence 143
Global Giants Have Successively Joined the Artificial Intelligence Camp 145
The Chinese Government Has Begun to Increase Investment in the Artificial Intelligence Field 146
China's Artificial Intelligence Companies Strive for the Upper Reaches and Increase Capital Investment 148
Bibliography 148

14 Tangible Hands 151
Establish a Coordinating Body for Overseeing Artificial Intelligence 152
Identify the Role of Different Levels of Regulatory Agencies 153
Strengthen Safety Controls 154
Minimize "Machine Bias" 155
"Who Encroached on My Privacy?" 155
Who Is Responsible? 156
The Importance of Anticipatory Regulation 158
Without Standards, Nothing Can Be Done 158

The Importance of Public Governance 159
China Should Incorporate Artificial Intelligence Regulations Into Strategic Considerations 160
Bibliography 161

15 **Kind AI** 163
Ethical Issues Become Artificial Intelligence's Most Formidable Challenge 164
Government and Organizational Strategies 166
Organizational Responses 169

16 **The Fight for Talent** 173
The Fight for Talent Has Fully Commenced 174
United States: Better Grasp the Needs of National Artificial Intelligence R&D Talent 174
Japan: Cultivating Teams of Professional Talent 175
UK: Full Scholarship Programs to Promote Science and Technology Education 176
China: A High Degree of Emphasis on Cultivation of Talent in Artificial Intelligence Domain 177
Whoever Obtains Artificial Intelligence Talents Obtains Everything Under Heaven 178
Bibliography 181

Part IV Law: Fairness and Justice in the Age of AI 183

17 **How To Be Accountable for AI?** 185
The Dilemma of Traditional Liability Theory: Can the Old Bottle Still Be Filled with New Wine? 185
Legislative Attempts in the Field of Autonomous Vehicles 186
Exploration of Legal Liability for Robots 188
Constructing a Reasonably Structured Liability System 190

18 **Deep Privacy Concerns** 191
Privacy and Data Protection Are Core AI Issues 191
Global Legislation on Privacy and Data Protection Is Heating Up 192

Challenge and Response: the Application of Anonymization Technology 195
Toward Legislative Dynamic Adjustments 197
Bibliography 198

19 Invisible Injustice 199
Algorithmic Decision-Making Is Increasingly Popular 200
Are Algorithms Fair by Default? 200
Algorithmic Discrimination Cannot Be Ignored 203
Discrimination in Crime Risk Assessment: Which Is More Reliable, Judges or Crime Risk Assessment Software? 204
Three Major Problems in Artificial Intelligence Decision-Making: Transparency, Accountability, and Fairness 206

20 Death of Authors 209
Are Artificial Intelligence Creations Protected by Copyright Law? 210
Does Artificial Intelligence Have Independent Intellectual Creative Abilities? 211
Do Artificial Intelligence Creations Meet the Threshold of Originality? 212
Other Intellectual Property Issues Relating to Artificial Intelligence 214
Bibliography 215

21 Who Am I? 217
Legal Personality for Artificial Intelligence Robots 218
Machine Rights 219
Who Will Empower Robots? 220
Which Rights to Give to Robots? 221
What Rights Can a Robot Have? 222
Robot Rights and Obligations 223
Bibliography 224

22 Ten Trends in Artificial Intelligence Law 225
Legal Practitioners Should Be Prepared for the Future 238

Part V Ethics: Human Values and Human-Machine Relations 241

23 Moral Machines 243
The Accelerated Arrival of Intelligent Machines 244
The Need for Moral Code 246
Realizing Moral Machines 247
Realizing Ethical and Moral Artificial Intelligence Requires a Comprehensive Governance Model 251
Bibliography 252

24 23 "Strong Regulations" for AI 253
Concerns Over Losing Control of Machines Are Long-Standing 254
Are Asimov's Three Laws of Robotics Reliable? 255
Exploring a New Round of "Strong Regulations" for Artificial Intelligence 257
The Future Requires "Controlling Spells" for Artificial Intelligence 260

25 The Future of Human-Machine Relations 263
Human-Machine Order in the Virtual World 263
Human-Machine Cooperation in the Technological Unemployment Crisis 264
Four Visions of Future Human-Machine Relations: Fantasy or Future Reality? 267
The Ultimate Question: Is Man a Machine? 271
Bibliography 272

Part VI Governance: Balanced Development and Regulation 273

26 From Internet Governance to AI Governance 275
From Management to Governance 275
Tracing Back to the Source of Internet Governance 276
The Expansion of Internet Governance 277
Bibliography 279

27 Challenges of AI Governance 281
Rules Lagging Behind Technology and Industry 281
Do We Really Understand the Technology? 282
The Ultimate Question: Walk Toward the World of AI, or Allow AI to Enter Our World? 283
Bibliography 284

28 AI Governance 285
Governance Should Be Established on the Foundation of Technological and Industrial Innovation 285
Moderate Regulation, Maintain the Humility of Authority 286
Don't Fall into the Trap of Over-generalized Safety Issues 287
Have the Promotion of Development and Innovation as the Goal 287
A Multi-level Governance Model That Encourages Multi-stakeholder Participation 288
Bibliography 289

Part VII The Future: Imagining the Future of AI Society 291

29 Whose Rice Bowl Has Been Smashed? 293
Hello, New Robot Colleague 293
"Artificial Intelligence +" Agriculture 294
"Artificial Intelligence +" Industry 295
Job Loss Warnings Are Sounding All-round 295
Who Will Be Replaced by Robots? 296
Robots Are Good Employees 299
But Are Robots Really Good Employees? 300
Fiscal Deficit and the Rise of Great Powers 301
Disappearing Iron Rice Bowl 302

30 War Robots 305
A New Round of Military Revolution and the Birth of Robots 305
R&D Trends in Autonomy and Intelligentization 307
The Sword of Damocles 309
Bibliography 312

31 Soulmate 313
A Master at Reading Minds 314
Know You Like "Her"? 315
Bibliography 316

32 New Productive Force 317
The Economic Revolution Driven by Artificial Intelligence 317
A New Round of "Apollo Missions" 318
Artificial Intelligence: New Factor of Production 319
Bibliography 321

PART I

Technology: The Reality of Disruptive Technology

The scope of the definition of Artificial Intelligence (AI) is a perpetual battle, and it is continuously updated by progress in this area. The currently popular "AI" is a very general concept that covers a broad range of different technologies under the two-letter abbreviation. Because the field of artificial intelligence has more than 60 years of history and is broad in scope, it is more complex and richer than most science and technology fields. How did artificial intelligence research begin? What stage has contemporary artificial intelligence research reached? What are the commonalities and differences between people's understanding and perceptions of artificial intelligence? In this part, we will take you through the past and present of artificial intelligence to reveal the truth of this disruptive technology.

CHAPTER 1

Gaps in Understanding of AI

Artificial Intelligence Rises Again

The year 2016 was a special year for artificial intelligence. At the beginning of the year, AlphaGo triumphed over Lee Sedol, a 9-dan rank player, and brought artificial intelligence technology, which has once again risen in the past decade, to the stage and into the public view.[1] In the past few years, technology giants have successively established artificial intelligence laboratories, invested more and more resources to capture the artificial intelligence market, and even completely transformed into artificial intelligence–driven companies, to intensively plan for the future of artificial intelligence. The Chinese government and other countries' governments all regard artificial intelligence as a strategic force driving the future, and are introducing strategic development plans and promoting overall progress from the national level to prepare for the artificial intelligence society that is about to arrive. This revolution will not be contained to laboratory research. At the same time, academic research and commercialization are promoting the conversion of artificial intelligence into products and services, so that the public can truly feel its existence. Especially in areas such as image and speech recognition, natural language processing, and other applications based on deep learning algorithms, they are rapidly becoming industrialized, and the racetrack has already been laid out.

[1] 9-dan rank is the highest rank for a professional Go player.

Tencent Research Institute et al. (eds.), *Artificial Intelligence*,
https://doi.org/10.1007/978-981-15-6548-9_1

Although we often talk about AI (artificial intelligence) in different settings, we have found that the "artificial intelligence" currently being hotly debated around the world is not exactly equivalent to the artificial intelligence defined by earlier academics. Artificial intelligence researchers and product designers, business people, policy makers, and the wider public generally use the term "artificial intelligence" in different contexts. On the other hand, like the previous terms "cloud computing," "big data," and "machine learning," the term "artificial intelligence" has been used by marketers and advertising copywriters without restraint. In the eyes of different groups, "artificial intelligence" seems to be both a panacea and a time bomb that causes massive unemployment.

As a technical term, "artificial intelligence" dates back to the 1950s. John McCarthy, an American computer scientist, and his colleagues proposed at the Dartmouth conference in 1956 that "letting machines achieve this type of behavior, that is, doing the same thing as humans," can be called artificial intelligence. In the following 60 years, artificial intelligence experienced "three ups and two downs," three times experiencing a rise, and two times falling into a valley. In addition to the continuous evolution of the direction of the technology itself, artificial intelligence has also gained many different levels of meaning due to the flexibility of interpretation. Before AlphaGo defeated Lee Sedol and Ke Jie, most people's impressions of artificial intelligence were limited to what they saw in the movies. For decades, a series of films such as *Artificial Intelligence*, *The Matrix*, *Her*, and *The Incredibles* have described humans' yearnings for and fears of "artificial intelligence." The concept of artificial intelligence is not only scientific common knowledge, but also a form of popular and commercial culture. The gap in understanding between a small group of AI experts and the public users of this "black box" technology is growing. So, during today's artificial intelligence renaissance, do we understand what AI means? What are its capabilities and limitations? Compared to the past, has the meaning of artificial intelligence changed?

Gaps in Understanding About AI

In order to assess people's understanding of artificial intelligence, Tencent Research Institute conducted an online survey from May to June 2017. Through the Tencent Questionnaire platform, we sent questionnaires to different groups directly or indirectly related to artificial intelligence, such as R&D personnel, technical personnel, product personnel, and researchers in the legal, policy, humanities, and social science domains. We received a total of 2968 responses from people from all walks of life. According to the survey data, the following questions were answered in turn: How do different groups of people understand and conceive of artificial intelligence and do differences exist? How do people accept and trust artificial intelligence in different areas? What issues need to be paid attention to in the process of artificial intelligence research? Are managers aware of the capabilities and limitations of artificial intelligence?

Among the surveyed respondents, the ratio of male to female was about 2:1, and the overall level of education was relatively high (see Table 1.1). Among them, 11.9 percent of people were engaged in occupations directly related to artificial intelligence, 45.7 percent of people were engaged in occupations indirectly related to artificial intelligence, and 42.4 percent of people were engaged in occupations unrelated to artificial intelligence. Practitioners directly or indirectly engaged in artificial intelligence included scientists, technical personnel, product and design staff, law and policy practitioners, humanities and social science researchers, media professionals, and entrepreneurs. We recognize the limitations of the above data and do not attempt to infer the overall situation in China.

The survey covered five important topics in the field of artificial intelligence: the understanding and recognition of artificial intelligence, future predictions about artificial intelligence, trust and acceptance of artificial intelligence, threat of artificial intelligence, and legal and research responsibilities for artificial intelligence.

Table 1.1 Composition of respondents

Variable		*Percentage*
Sex	Male	67.5
	Female	32.5
Education level	Undergraduate	50.7
	Master's	37.2
	PhD	3.8

Understanding and Recognition of Artificial Intelligence

This section mainly analyzes people's impressions and understanding of artificial intelligence. In the eyes of the people, what does artificial intelligence mean? In this survey, we did not predefine artificial intelligence in a general or in a narrow sense. Instead, we asked the public a wide range of questions about their first impression of artificial intelligence, their understanding of existing achievements, and their imaginings of the future.

Impressions of AI: When "Artificial Intelligence" Is Mentioned, What Do You First Think Of?

More than half of the respondents mentioned AlphaGo and robots. Common words also included "self-driving cars," "terminator," "Siri," and "big data." When talking about artificial intelligence, people often confused it with the concept of robots. But the current wave of artificial intelligence is more about the flourishing of deep learning algorithms based on big data. It cannot be equated with previous attempts to create "Artificial General Intelligence" restoring robot forms with human intelligence and behavior.

What Capabilities Does the AI Already Have?

"Artificial intelligence" is a collective term for a group of technologies. To understand the capabilities already possessed by AI, it is necessary to understand the development of artificial intelligence in the current technical field and the problems it can solve, instead of seeing artificial intelligence as a type of general ability. For example, decision-making capabilities involve reinforcement learning. Creativity refers to generative models associated with creation, and will have very good applications in the field of content creation. Affective computing research is trying to create a type of computing system that can perceive, recognize, and understand human emotions, and can make intelligent, sensitive, and friendly responses to human emotions, that is, giving the computer the ability to observe, understand, and generate various emotional characteristics like humans. At present, relevant research has made some progress in facial expression, posture analysis, and emotion and recognition aspects of speech. The machine understands your emotions, but this does not mean that it will have "empathy" like humans.

Future Predictions About Artificial Intelligence

Artificial intelligence has entered a period of rapid development. With the application of artificial intelligence blossoming in all sectors, a society in which human beings and artificial intelligence co-exist harmoniously with each other is nearing ever closer. What kind of artificial intelligence society will we usher in?

Will AI Be Commonplace in Society After Ten Years?

Overall, 47.8 percent of respondents believed that artificial intelligence will be commonplace after ten years. After further analysis, the closer a respondent's connection to artificial intelligence, the more likely they were to believe that artificial intelligence would be commonplace in the next ten years.

Will Artificial Intelligence Have a Positive Impact?

The survey results revealed that respondents were optimistic about the impact of artificial intelligence on society.

The more knowledgeable a respondent was about artificial intelligence, the more likely they were to believe that artificial intelligence will have a positive effect.

Among the interviewees who identified as being "very knowledgeable" about AI, 82.63 percent agreed that artificial intelligence will have a positive impact on society; among respondents who chose "not much understanding" of artificial intelligence, only 59.30 percent believed that artificial intelligence will have a positive effect on society. Among respondents who had used artificial intelligence products, 73.38 percent thought that artificial intelligence will have a positive impact on society. Among respondents who had not used artificial intelligence products, the figure was 9.1 percentage points lower, at 64.28 percent. Lack of understanding and even misunderstanding of artificial intelligence may cause people to fall into an "ignorant fear" of artificial intelligence.

Will Artificial Intelligence Develop Consciousness?

Consciousness is the most magical mental ability of human beings, and it is also a very mysterious and complex phenomenon. Since the 1990s, many philosophers, psychologists, and neuroscientists have started to study what is known as "machine consciousness." There are two distinct viewpoints on the existence of phenomenal consciousness.[2] One is the mystical view that the only thing common to our neurobiological systems is the subjective experience. This kind of phenomenal consciousness cannot be reverted to a physical mechanism or logical description. It cannot be grasped by the human mind. The other is an eliminativist view that the machine is just a zombie. In fact, there is nothing but machines, objects that cannot have any subjective experience. The argument for machine intelligence itself includes people's different understanding of consciousness. For the ultimate goal of achieving artificial intelligence, consciousness is a problem that cannot go unnoticed. If "artificial general intelligence" becomes possible in the future, it must be accompanied by the emergence of "machine consciousness." For the current round of artificial intelligence based on machine learning, this is still a relatively distant research direction.

AI Trust and Acceptance

Acceptability is the key to implementing artificial intelligence. User trust in artificial intelligence systems is the prerequisite for artificial intelligence systems to produce social benefits. Secure and steady trust requires constant trial and error. Trust requires a system of practice that helps guide the security and ethical management of artificial intelligence systems. This includes coordinating social norms and values, algorithmic accountability, compliance with prevailing legal norms, as well as ensuring the integrity of data algorithms and systems, and protecting personal privacy.

[2] Phenomenal consciousness is the felt, subjective, or "what it's like" aspect of mental states (see Nagel 1974).

In Which Areas Do You Want to Use Artificial Intelligence?

The results of the survey showed that respondents wanted to use artificial intelligence in smart homes, transportation, elderly/child care, and personalized recommendations.

Level of Acceptance in Nine Major Fields: Are We Prepared?

According to observations of the current landscape of enterprises and research in artificial intelligence, we selected nine common application areas: autonomous driving, virtual assistants, research/education, financial services, medicine and diagnostics, design and artistic creation, legal practices such as contracts and lawsuits, social companionship, and services and industry. Respondents were asked to answer regarding each of the nine areas to what extent they could be handed over to "artificial intelligence": (1) humans do it themselves; (2) mainly human beings, with support from artificial intelligence; (3) mainly artificial intelligence, with human supervision; (4) artificial intelligence replaces people; (5) unclear.

Areas of high acceptance of artificial intelligence include: services and industry, autonomous driving, financial services, and virtual assistants. Forty-two percent, 41 percent, 41 percent, and 40 percent of respondents, respectively, believed that artificial intelligence should be doing the bulk of the work, with human supervision. Especially in services and industry, 40 percent of respondents believed that artificial intelligence can replace people.

Areas where AI acceptance is relatively low include: research/education, medical and diagnostics, social companionship, and legal practices such as contracts and lawsuits. Fifty-seven percent, 49 percent, 43 percent, and 39 percent of respondents, respectively, believed that humans should do the bulk of work in these fields, supported by artificial intelligence.

The lowest level of artificial intelligence acceptance is in the field of design and artistic creation. Forty-seven percent of respondents believed that humans should perform the tasks in these fields themselves. Only 4 percent of respondents believed that artificial intelligence could replace people in this field.

According to people's answers to the above questions, it is easy to draw a conclusion that meets the beliefs of the public; that is, the higher the degree of mechanization in the work, the more people want it to be completed by artificial intelligence. For work that requires creativity, people are more confident in human capabilities.

The fact is that, unlike the previous wave of automation that only affected mechanical labor, artificial intelligence has increasingly appeared in research and art. At the end of 2016, Sony released a pop song "Daddy's Car" created by artificial intelligence. The track was created by Flow Machines, a Sony Computer Science Laboratory artificial intelligence program, and involved discovering special style by analyzing a database with a large number of songs. Artificial intelligence has been able to create poems and songs. In the field of art and creation that people thought could not be replaced by machines, the trend of man-machine integration has gradually emerged. However, Flow Machines head Pacht said that although artificial intelligence can now create "perfect" songs, only musicians can create unique works.

AI Interaction Mode: "Natural Language Communication" Has Become the Preferred Mode of Human-Machine Interaction

Every technological revolution drives the evolution of methods of interaction at the same time. With the rapid development of language recognition technology and natural language processing technology (NLP), speech recognition has gradually become a common interaction mode of intelligent machines. Some analysts estimated in a report that by 2020 dialogue between ordinary people and machines will exceed dialogue between spouses. The report does not indicate whether the reason is the increase in dependence on AI technology or the deterioration of future spousal relationships, but it may also be a combination of the two factors. At present, the transition from "screen operation" to "chat interface" in electronic devices has become a trend. A number of players and products have emerged in the field of voice interaction. There are Amazon Alexa, Google's Google Assistant, Tencent Cloud's Xiaowei, and Baidu's Duer. These products use dialogue as an interactive method to control different smart devices. All technology companies are accelerating this transition and are striving to enter the next generation of artificial intelligence services.

The Threat of AI

When artificial intelligence gains territory in various fields, there are also various hidden concerns about AI. Some people worry that AI will largely replace human labor. Some people worry that the development of AI will not be controlled. Films such as *Metropolis* and *Terminator* have such arguments and they express a kind of fear. When a strong artificial intelligence system is created, perhaps its wisdom will be far beyond that of humanity, bringing some unimaginable risks. How should one understand the threat of artificial intelligence?

Is It Possible for Artificial Intelligence to Control Humans?

Among the people who said that they did not understand artificial intelligence, 38.47 percent believed that artificial intelligence might control humans. Among the people who said they had some knowledge or were extremely knowledgeable, this ratio was 36.76 percent and 27.8 percent, respectively.

When Will "Strong Artificial Intelligence" Come?

Regarding the threat of artificial intelligence, the most famous is the "AI threat theory" promulgated by Elon Musk. He has publicly stated on several occasions that artificial intelligence may become the greatest threat to human civilization and called for the government to quickly adopt measures to supervise this technology. Contrary to Musk's "AI threat theory," many artificial intelligence industry experts and scholars including Zuckerberg, Kai-fu Lee, and Andrew Ng have all expressed a view that the artificial intelligence threat to human survival is still far away. The biggest difference between the two sides on whether artificial intelligence threatens humans comes from the different understanding of "artificial intelligence." The "artificial intelligence" in the context of what Musk is mainly referring to is "strong artificial intelligence" (or "artificial general intelligence"), that is, having the ability to handle multiple types of tasks and adapt to unforeseen circumstances and capabilities. Zuckerberg's term "artificial intelligence" refers to artificial intelligence in a narrowly defined professional field. At present, there is no conclusive scientific community consensus on when "strong artificial intelligence" will be achieved. More than half of scientists and technical researchers believe that "strong artificial intelligence" will not be realized before 2045, while non-technical groups predict that it will be realized in a shorter time.

Legal and Research Responsibilities Relating to AI

The civilization we love can be said to be the product of intelligence, so the use of artificial intelligence to amplify human intelligence has the potential to bring unprecedented prosperity. Of course, we must develop technology on the premise of benefiting humanity.

The process of continuous development of artificial intelligence raises new questions about ethics and legal liability. For example, who should have legal liability when artificial intelligence systems pose a potential threat to users?

(1) When artificial intelligence causes damage to human life and property in the fields of autonomous driving, medicine, and so on, who do you think will bear the legal responsibility?
(2) From which stage should we consider ethical, legal, and social influences?

Only 1 percent of respondents believe that there is no need to consider the ethical, legal, and social impacts of artificial intelligence (1.2 percent chose "Don't Know"), but different groups have different considerations about when to start considering these issues.

Compared with the scientific research community, humanities and legal groups pay attention to the ethical, legal, and social impacts of artificial intelligence at an earlier stage of artificial intelligence. Humanities and social science researchers and policy and legal groups believe that ethical, legal, and social influences should be considered from the basic research phase of artificial intelligence. Scientists, entrepreneurs/entrepreneurs, and technicians consider the ethical, legal, and social impacts of artificial intelligence relatively late.

Misconceptions About AI

Based on the above research, we have listed the following seven common misconceptions in the field of artificial intelligence:

Misconception 1: Artificial intelligence equals robots.

Fact: "Artificial intelligence" is a term that encompasses a large number of subfields and covers a wide range of applications.

Misconception 2: The benchmarks for artificial intelligence are specific verticals like Online-to-Offline (O2O), e-commerce, and consumer upgrades.

Fact: Artificial intelligence provides technical tools for the upgrading of the entire industry.

Misconception 3: Artificial intelligence products are distant from ordinary people.

Fact: In real life, we are already using AI technology and it is everywhere. For example: email filtering, personalized recommendations, WeChat's voice-to-text, Apple's Siri, Google's search engine, machine translation, autonomous driving, and more.

Misconception 4: Artificial intelligence is one technology.

Fact: AI contains many technologies. In a specific context, if a system has one or several capabilities in speech recognition, image recognition, retrieval, natural language processing, machine translation, and machine learning, then we think it has a certain type of artificial intelligence.

Misconception 5: Artificial general intelligence will come in the short term.

Fact: In the short term, artificial general intelligence is not the mainstream research direction in the industry. We are more likely to see thorough penetration of deep learning in all areas.

Misconception 6: Artificial intelligence can generate consciousness independently and autonomously.

Fact: Current artificial intelligence is still some distance away from artificial general intelligence. Tool-based artificial intelligence cannot generate awareness.

Misconception 7: Artificial intelligence will replace human labor in the short term.

Fact: The application maturity of artificial intelligence varies greatly in different fields. Although artificial intelligence can already overcome the strongest professional player in the world in the field of Go, it may take 50 years to create its own best-selling works. Tool-based artificial intelligence and human capabilities are complementary in many contexts. In the short term, human-machine collaboration is more likely to occur.

Welcoming the Future

The resurgence of artificial intelligence is not accidental. The reason why this round of artificial intelligence can flourish is that we have ample amounts of data, powerful computing resources, and more advanced algorithms. The new generation of changes have an important feature: deep learning based on big data. In 2006, the basic theoretical framework for deep learning (deep neural networks) was validated, and as a result, artificial intelligence opened up a new round of flourishing. In 2010, the first breakthroughs were made in the field of speech and natural language processing. Since 2011, deep learning has surpassed humans in the field of image recognition, this type of algorithm has shined in various fields. The changes in the artificial intelligence that the industry talks about in all walks of life are also centered on deep learning and a series of related data processing technologies.

We are now only in the initial phase of this wave of artificial intelligence. Rejecting the external publicity, we need to concretely and more accurately understand artificial intelligence. In the remaining chapters of Part I, we will unveil the past and present of artificial intelligence and the business and social changes it is bringing. This book is divided into seven parts: Technology; Industry; Strategy; Law; Ethics; Governance; and Future. The authors of this book also cover different research subjects. Layer by layer, the authors have deconstructed the concept of artificial intelligence and its development path from different perspectives, giving you an appreciation of the bumpy challenges and bright spots of artificial intelligence. Artificial intelligence will eventually reshape this world. These trends have now been observed in all walks of life. At the same time, every breakthrough in artificial intelligence will bring ethical and legal challenges. We will rapidly research the ethical, legal, and social impacts of artificial intelligence in the early stage of technological development and welcome a society of "human-machine coexistence."

Welcome to the new world of artificial intelligence.

Bibliography

"Top Strategic Predictions for 2017 and Beyond: Surviving the Storm Winds of Digital Disruption". *Gartner*. 14 October 2016. 23 March 2020. https://www.gartner.com/en/documents/3471568.

Zhou, Changle. "机器意识能走多远:未来的人工智能哲学". *民论坛·学术前沿*. 13 (2016): 81–95.

CHAPTER 2

Artificial Intelligence's Past

When AI comes up, we think of scenes from films and television: the AI operating system Samantha from *Her* trapping a person into loving it, the blow-up medical robot Baymax from *Big Hero 6*, robot receptionists from *Westworld* that wonder through the park gradually gaining consciousness. All of these represent the great expectations we have for artificial intelligence.

Going back in time to the summer of 1956 when the Dartmouth Summer Research Project on Artificial Intelligence saw John McCarthy, Marvin Minsky, Nathaniel Rochester, and Claude Shannon, along with over six other scientists, together discus unresolved questions in computer science and for the first time put forward the idea of artificial intelligence (AI). After this meeting, began the first AI spring, but hampered by hardware and software conditions at the time, AI was restricted to simulating human brain function and researchers could only work on specific issues in a few fields to create some theorem proving machines, draughts and chess procedures and robots that played with toy bricks. In that time when computers could only do numerical calculations, such uses could only reveal a small part of AI's applications and became seen as reflecting what AI is.

Moving into the twenty-first century, the arrival of deep learning has set off another surge in AI. An endless stream of applications as small as your phone's Apple Siri to intelligent security systems have appeared in academic papers, in the news, and in people's daily lives. Amid these, one that stands out as a milestone is DeepMind-developed AlphaGo beating top-level professional Go player Lee Sedol four games to one in a battle of

Tencent Research Institute et al. (eds.), *Artificial Intelligence*,
https://doi.org/10.1007/978-981-15-6548-9_2

man-versus-machine to become the world champion at Go. Even people who knew nothing about AI finally began to appreciate its power.

Although AI technologies have developed rapidly in recent years, it is not easy to clearly define AI. AI research is normally seen as a new technical science made up of theories, methods, technologies, and applications that can simulate, extend, and expand human intelligence. Lots of things in the everyday life of people, such as calculation, observation, dialogue and learning, all require "intelligence." "Intelligence" can supervise the forecasting of stock prices, has the ability to understand the contents of a picture or video, or can communicate between people in writing or via language; it can create a store of information that continuously self-improves, paint, write poems, drive a car or fly an airplane. In our imaginations, if a machine can carry out one or many of these tasks, then it can be seen as having a degree of "artificial intelligence." Today, the connotations of AI have expanded greatly. It is a cross-disciplinary science that covers many fields including computer science, statistics, neurology, and social sciences. People hope that via the research of AI, human intelligence can by simulated and expanded, to assist or even replace many capabilities of humans, including distinction, cognition, analysis, and decision-making.

Layers of AI

If you want to systematically describe AI, starting from the lowest layer and working upward, the bottom is the basic infrastructure layer, then the algorithm layer, then the technical layer and the application layer. The basic infrastructure includes the hardware/computing power and big data; the algorithm layer includes various machine learning and deep learning algorithms; further up are various technical aspects, such as computer vision or language technologies with cognition/analysis functions, natural language processing (NLP) technology that provide understanding/assessment capacity, technologies that provide decision/interactive capacities for planning and decision-making systems and for big data/statistical analysis. Every technical trend is not only a specific technology; the highest layer is industry solutions, with currently rather mature fields being in finance, security, transport, medicine, and games.

Infrastructure Layer

Looking back upon the development of AI, every development in basic infrastructure made a clear improvement to the computing and technical layers. Computers came into being in the 1970s and in the 1980s they became widespread; then in the 1990s the computation speed and storage capacity of computers grew and the rise of the Internet brought the electrification of data, all of which worked to greatly improve AI. And when it came to the twenty-first century, the effectiveness of these improvements became even more striking. The appearance of large-scale Internet service providers, the accumulation of big data that came from search and e-commerce and the upgrading of graphics processing units, or GPUs, and heterogeneous/low power consumption microchips brought about the emergence of deep learning and set alight this explosive wave of AI.

In this wave, the role of an explosive growth in data cannot be ignored. We know that the volume of training data is an important fuel for AI, and the scale and abundance of data is especially important. If we look upon AI as a newborn baby, the volume, specificity, and depth of a particular field's data are the infant formula used to nurse this potential genius. The quantity of the infant formula determines whether or not the baby grows, but the quality of the formula determines the baby's future level of intellectual development. Since 2000, thanks to the Internet, social media, mobile devices, and sensors becoming universal, the store of data created across the world has increased. According to a report by the International Data Corporation, or IDC, the total data in the world is estimated to be over 40 zettabytes by 2020 (that's the equivalent of forty trillion gigabytes). This is 22 times the amount of data in 2011. In recent years, the amount of data globally has grown at a rate of 58 percent annually, and the future speed will be faster still. Compared to the past, the amount of information stored in data today and its dimensions are both ever greater, from simple text, pictures, and sound data, to motion, attitude, and trajectory data for human behavior, and on to environment data like geolocation and temperature. As the scale of data increases and it becomes more abundant, its uses for modelling will as a matter of course also become greater.

Another aspect is the improvement in computing power, which has also had a clear effect. The appearance of AI chips has clearly raised the speed of data processing, and is in particular clearly preferable to a traditional CPUs for massive data processing. From this foundation of CPUs that, though skilled in processing/control and complex processes, remained high power consuming, there emerged skilled parallel computing GPUs, as well as FPGA (Field Programmable Gate Array) and ASIC (Application Specific Integrated Circuit) that were more suited to deep learning models and had good operating efficiency. The power consumption of chips has also become ever lower, and their flexibility is ever higher, to the extent that they now measure up to carrying out special function deep learning algorithms.

Algorithm Layer

When it comes to the algorithm layer, we must first clarify some concepts. So-called machine learning is using algorithms to allow computers to mine information from data like a human can; "deep learning" is a subset of "machine learning," which, relative to other learning methods, has wider parameters and more complicated models, allowing the understanding of the data to be deeper and more intelligent. Traditional machine learning takes place step-by-step, and the optimization of each individual step does not necessarily lead to the optimization of the end result. Another point for machine learning is that the manual selection of features takes time and energy, plus it requires expertise, meaning that it is to a great extent dependent on experience and luck. But deep learning starts from the original features and automatically learns how to combine them with the higher-level features. The whole process is self-contained end-to-end and directly ensures that the end result will be optimized. However, the middle layer is a black box, and we don't know what features the computer has extracted.

There are some typical problems that are hit upon in machine learning. The first is the problem of unsupervised learning: a fixed amount of data is provided, and information is discovered from the data. The inputs do not have dimensional labels from historic data, and the task's outputs are assembled and classified data. For example, asking the machine to automatically classify particular types of fruit together from a particular basket. How does the machine do it? First, it finds the vectors of each particular fruit, such as color, smell, and shape. Then group fruits with similar vectors (vectors that are relatively close) together: red, sweet, round is one kind; and yellow, sweet, and long is another. When people come looking, they will see that the first kind is apples, the second kind is bananas. This is unsupervised learning, and typical examples are user clustering or news clustering.

The second kind of problem is supervised learning: from a set of data, infer this data's labels. The inputs are historic data with dimensional labels, and desired outputs are the inference model that is being used. For example, the set is a basket of fruit, in which each fruit has a fruit name label. You ask the machine to learn from those labels, and then to predict the names of new fruit. The machine will still have to look at vector expression, and via the labels for a fruit with a particular name, it will discover that red, sweet round corresponds to apples, while yellow, sweet, and elongated corresponds to bananas. As such, faced with new fruits, the machine can use the vectors to know whether it is an apple or a banana. Supervised learning is typically used to for recommendation or forecasting related questions.

The third kind is reinforcement learning problem: from a set of data, chose the action that will maximize long-term reward. The input is historic conditions, actions and corresponding rewards, and the desired output is the optimum action for the current situation. The difference between this kind and the former two kinds is that reinforcement learning is a process for learning trends, but does not have a set learning goal, and there is no precise measurement criterion for outcomes. Reinforcement learning is a sequential decision-making problem where a computer consecutively chooses some actions, and without any dimensional labels telling the computer what it should do, the computer must first try out some actions, see what the results are, then, via judging whether those results were right or wrong, provide feedback on its former actions. To illustrate with an example, suppose that at lunchtime you want to go downstairs to eat lunch, and you have already tried out a portion of those restaurants near you, but not the whole lot, you can then choose the best possible restaurant from those that you have already tried (this is called exploitation), or you can try a new restaurant (this is called exploration). The latter option could allow you to discover an even better restaurant, or it could allow you to eat at an unsatisfactory restaurant. And when you have already tried enough restaurants, you can summarize your experience (restaurants with high scores on Dazhong Dianping are normally not bad; those restaurants near the office are not usually as good as those further away); these experiences will help you find more reliable restaurants. Many problems of manipulating kinds of strategic decision-making are reinforcement learning problems, such as asking machines to fly an unmanned drone stably via various parameter adjustments, or winning points in a video game via using a variety of button combinations.

An important subdivision of machine learning algorithms is neural network learning. Although it took until the twenty-first century to become familiar to people due to the victory of AlphaGo, the history of neutral network goes back to at least 60 years ago. In these 60 years, neutral networking has gone through high and low points, due to various restrictions of data, hardware, and computing power at the time, and it would again and again hit bottlenecks and be left out in the cold, before a new breakthrough would bring it back to people's attention. The latest being the close attention that has come with the rise of deep learning.

From the 1940s, there have been scholars undertaking neural network research. McCulloch and Pitts published "A Logical Calculus of the Ideas Immanent in Nervous Activity," which was seen as the first essay on neural networks. Neural psychologist Herb published a book titled *The Organization of Behavior*, putting forward what people came to call "Hebb's principles" of machine learning. The first big breakthrough came in 1958, when Rosenblatt used a computer to simulate a model he invented called "perceptron," which could handle simple visual processing tasks and was a rudimentary support vector machine (a kind of algorithm that could quickly and reliably classify) for neutral networks. At this time, this model of algorithm that simulated the human brain was widely flattered, and government organizations including defense departments one after another started to support neural network research. The craze of neural networks continued for over ten years; then in 1969, Minsky and others proved that the perceptron had capacity restrictions for foundational reasoning problems like resolving XOr, or "exclusive or", problems (a classification problem) and this flaw extinguished the enthusiasm that people had had for neural networks, and government funding gradually stopped, which brought about the ten-year neural network "AI winter." At this time, Werbos in 1974 proved that neural network algorithms could effectively resolve XOr problems if they added an additional layer and used back-propagation (BP), but due to neural networks still being in "low tide" at the time, the finding did not garner much attention.

It wasn't until the 1980s that neural networks eventually recovered. Physician Hopfield in 1982 and 1984 published two treatises on artificial neural networks and proposed a new form of neural network that could resolve a large portion of pattern recognition problems, as well as being able to provide an approximate solution to one kind of combinatorial optimization problem. His research brought about a huge response, and people again came to recognize the power and real-world applications of neural networks. In 1985, Rumelhart, Hinton and many more neural network scholars successfully realized and used back-propagation (BP) algorithms to train neural networks, and for a time BP became a dedicated algorithm for neural networks. After this, ever more research started to emerge. In 1995 Yann LeCun and others inspired by a model of biological vision changed to Convolutional Neural Networks (CNN). These networks simulated the cells of the visual cortex (with a small number of cells having a special sensitive visual region and individual neurons responding only when an edge with a particular direction exists). They used a similar means of computation to carry out image classification tasks (via searching for the lowest level of simple characteristics, such as edges and curves, then using a series of convolutional layers to establish an abstract idea of the image), and were able to achieve the best results of the time for small-scale problems such as distinguishing between handwriting styles. After 2000, Bengio and others pioneered the development of established language neural network models.

Until 2001, when Hochreiter and others discovered how to use the back-propagation algorithms, neural network algorithms would degrade after each unit was saturated; even if the training model could easily find a fit after a number of data iterations, the distributions of training collection and the test collection data would not match (just like taking a test at school, some people take the approach of attempting to memorize a sea of exam questions in each topic. But as soon as the topics change, even slightly, they can't do the test. Because the machines use a very complicated means of remembering every question, they don't have rules for abstract generalizations). Neural networks were again abandoned. Yet, neural networks still did not go quiet, and many scholars continued to untiringly carry out research. In 2006, Hinton and other students published a paper in *Science* magazine, sparking a surge in interest in Deep Learning. Deep Learning can find complicated structures within big data, and it can greatly elevate the efficiency of neural networks. From 2009 onward, Microsoft Research and Hinton worked together to develop neural networks capable of speech recognition with a mistaken recognition rate of under 25 percent. In 2012, Hinton again led students to reach an astonishing result in classification of images from the largest image data store ImageNet with Top-5 error rate being reduced from 26 to 15 percent. Afterward the next symbolic time was in 2014, when Ian Goodfellow and others released an academic paper with the title "Producing Opposition Networks," marking the birth of GANs, and from 2016 the idea became a force to be reckoned with in the academic and business worlds, as well as a strong algorithmic framework for the establishment of the unsupervised learning model. Today, after neural networks have gone through ebbs and flows, they are again are on the frontlines, with its influence being seen everywhere from image recognition, to speak recognition, to machine translation.

In addition, other shallow learning algorithms are also continuing to develop along another path and are even replacing neural networks as the most favored algorithm. Up to this day, even with neural networks at their apex, these shallow learning algorithms occupy certain niches for some tasks. In 1984, Brennan and Friedman put forward the decision tree algorithm as a calculation model, which represents a way of shining a light on the relationship between the target property and the target value. In 1995, Vapnik and Cortes put forward a supportive support vector machine (SVM) that used a classification hyperplane to separate samples and efficient classify. This kind of supervised study method made wide-scale use of statistic classification and regression analysis. In view of the power of SVM's theoretical position and actual results, machine learning research can from now on be slipped into two teams of neural networks and SVM. In 1997, Freund and Schapire put forward another robust ML model, AdaBoost, the largest distinguishing feature of which is to combine weak classification capacity with strong classification capacity, which has widespread application in facial recognition and testing. In 2001, Brieman suggested that many decision trees could be combined into a random forest, and can handle a large volume of input variables with a quick learning process and a high degree of accuracy. With the advancing of this method, SVM became more efficient in a number of tasks that neural networks had previously held, and neural networks could not compete with SVM. Later, although deep learning's artisan gave neural networks a second spring and allowed it to achieve leading results in image, voice, and NLP spheres, this was not the end of the story for other kinds of machine learning. The problems of training costs of deep neural networks and the complexity of transferring references continue to draw castigation, and SVM's simplicity allows it to maintain niches of widespread use in text processing, image processing, web search, financial credit reporting and other spheres.

Another important area is reinforcement learning, which is a concept that became well known due to AlphaGo. From its birth in the 1960s, it has slowly developed without fanfare until its innovative combination with deep learning in AlphaGo gave it a new lease of life.

In 1967, Samuel-invented chess playing process was the earliest rudimentary application of reinforcement learning. But in the 1960s and 1970s, people confused reinforcement learning research with problems like supervised learning and pattern recognition, causing its development to be slowed. In the 1980s, along with the improvement of infrastructure and neural network research, reinforcement learning research again reached a climax. In 1983, Barto, via reinforcement learning, was able to sustain a handstand for a relatively long period of time. Another outstanding reinforcement learning scholar Sutton also advanced a number of important reinforcement learning algorithms, including the Adaptive Heuristic Critic algorithm he put out in 1984 and afterward, in 1988, the temporal difference method. In 1989, Watkins put forward the famous Q-learning algorithm. As a number of important algorithms were put forward, when we reached the 1990s, reinforcement learning had gradually become an important component of machine learning.

The most recent milestone came in 2016 when David Silver of the company DeepMind, which is owned by Google, innovatively combined deep learning and reinforcement learning, creating the AlphaGo software, which successively beat Li Shishi, Ke Zhi and others to become the world Go champion, demonstrating the tremendous power of reinforcement learning.

Technological Developments

Computer Vision

"Seeing" is a capacity people and animals all have. A just born baby only needs a few days' time to begin imitating its parents' expressions; people can find the focal point in pictures with complicate structures and can recognize a familiar person in a dark environment. Alongside the development of AI, machines have also tried to match or even outdo humans in this capability. The history of computer vision can be traced back to 1966 when AI scholar Minsky was setting his students' homework and asked them to edit a process to allow a computer to tell us what we are seeing via a video camera, which is thought to be the first task description for computer vision. Through the 1970s and 1980s, with the arrival of modern computers, computer vision technology has also begun to grow. People have begun to try to get computers to say what it is seeing; thus, they first had to think how humans saw things as a reference point. The first lesson drawn was something that people at the time universally believed, that people can see and understand things because they can observe those things using their two eyes. As such, to let computers see images, they first had to find a way for them to recover three-dimensional things from two-dimensional images; this is what is known as "three-dimensional reconstruction." The second lesson drawn is that people believe that the way that they are able to recognize an apple is because people have a priori knowledge of apples, such as that an apple is red, round and smooth. If you can build up a knowledge store for a machine and match that images to the stored knowledge, can you then let machines recognize and even understand the objects it sees? This is the so-called a priori knowledge store method. This period of application was mainly optical character recognition (OCR), workpiece recognition, micro/navigation picture recognition and so on.

When the 1990s arrived, computer vision technology had achieved some major developments and had begun to be widely used in industry. One aspect was that graphics processing unit (GPU) and digital signal processing (DSP) led to rapid progress in image processing hardware; another aspect was that people started to try out different algorithms, including the integration of statistical methods and local feature descriptions. In the "a priori knowledge" method, an object's form, color, surface grain and other characteristics can change from different angles, in different lighting, with different cover and due to different angles of perception and observation environments. As such, people found a means of relatively accurate matching even if the angle of perception or the observation environment changed, which was to use recognition of local features to judge objects and to create an index of objects' local features.

Entering the twenty-first century, profiting from the sea of data that came from the rise of the Internet and digital cameras, machine learning methods became more widely applied, and computer vision development sped up. Former processing methods based on many foundational rules were replaced by machine learning that automatically summed up objects' special characteristics from a sea of data and then carried out recognition and determination. This period saw a lot of applications spring up, including typical camera facial examination, facial recognition for security and defense, car license plate recognition and so on. Data's accumulation also gave birth to many data appraisal and measure collections, such as the authoritative facial recognition and facial comparison recognition platforms FDDB and LFW, with the most influential being ImageNet, which contains 14 million labelled images divided into over 10,000 categories.

After 2010, with the help of deep learning power, computer vision technology achieved explosive growth and was industrialized. Via deep neural networks, the perception precision in various tasks related to vision all achieved substantial improvements. In the most authoritative global computer vision competition ImageNet Large Scale Visual Recognition (ILSVR), the Top-5 error rate for thousand category object recognition was 28.2 and 25.8 in 2010 and 2011 respectively; then from 2012 and with the introduction of deep learning, the rates for the next four years were 16.4, 11.7, 6.7 and 3.7, demonstrating a clear breakthrough. Due to the efficiency gains, computer vision technology applications quickly expanded; apart from the relatively mature field of security and defense, they were also used for facial recognition and identification verification in the financial sphere, for searching photographs of goods in e-commerce, for intelligent image diagnosis in medicine, as a visual input system for robots and unmanned cars, and in many interesting fields including automatic image classification (image recognition and classification), image description generation (image recognition and understanding) and so on.

Language interchange is humankind's most direct and succinct means of communication. For a long time, making machines able to "hear" and "speak" and carry out an unimpeded exchange with humans has been a consistent dream of the AI and human-computer interaction fields. Even before the arrival of electric computing, people had the dream of making computers capable of recognizing speech. The "Radio Rex" toy dog created in 1920 was perhaps the world's first speech recognition machine, when you shouted "Rex," the dog would pop up from the floor. But in actual fact the technology that it used was not really speech recognition, but was rather a spring that would automatically release when hit with 500 hertz of noise, and 500 hertz was the first resonance peak when people shouted "Rex." The first real electric computer–based speech recognition system came out in 1952, with AT&T Bell Labs developing a speech recognition system called Audrey that would tell apart ten English words with an accuracy rate of 98 percent. In the 1970s, large-scale speech recognition research began, but at the time the technology was only in its early stages and remained at the level of distinguishing between individual terms and sentences using a limited vocabulary.

The 1980s were an era of technological breakthroughs, an important reason being that the global telex work accumulated a large volume of text, and these texts were machine-readable and could be used for matrixes as statistics and training. The focus of research also gradually turned to larger vocabulary and continuous speech recognition from non-specific individuals. The most important change at the time came from the statistics-based mode of thinking replacing the matching-based mode, with a key bit of progress being the Hidden Markov Model (HMM) theories and applications trending toward improvements. The industrial world began to make widespread use of the technology, and Texas Instruments researched and developed a phonetics learning machine called Speak&Spell, speech recognition service trader Speech Works was set up and the US Defense Department's high-level research planning department (DARPA) also assisted and supported a series of speech-related programs.

Speech recognition was essentially mature by the 1990s, and the main framework of Gaussian Mixture Model (GMM-HMM) was essentially stable, but there was still a gap between the recognition results and actual applications and the progresses of speech recognition research gradually eased. Due to the fervor around neural network technologies in the 1980s and 1990s, neural network technologies were also used for speech recognition, and various combinational models of layered perception machines and HMM were put forward Multilayer Perceptron (MLP-HMM). But the functionality of these models could not outdo the GMM-HMM framework. Breakthroughs brought about the appearance of deep learning. As deep neural networks were applied to speech acoustic modelling, people one after another achieved breakthroughs in phoneme distinction tasks and large vocabulary speech recognition tasks. The GMM-HMM speech recognition framework was replaced by the DNN-HMM speech recognition system, and with the continuous improvements to the system, the recognition results from the use of a revolving neural network arose, the LSTM, and many speech recognition tasks (especially close range) reached a standard where they could be used in people's daily lives. Therefore, the intelligent speech-activated assistants such as Apple Siri, and the leading intelligent hardware Echo and so on could be used. And the universalization of these applications again further expanded collection channels for language data resources, creating an abundant fuel store for language and acoustic model training, and making it possible to set up large-scale models that make use of language and acoustics.

Natural Language Processing

In people's daily lives, language exchanges are an important channel for individuals to exchange information and communicate. As such, for machines, being able to naturally communicate with humans and to understand the meaning of human expressions and make suitable responses is considered an important reference for the measurement of intelligence level; natural speech processing has thus become an unavoidable subject.

In the 1920s and 1950s, along with the appearance of electronic computing, many natural language processing tasks were attempted, including the most common task of machine translation. At the time there were two methods of natural language processing: the symbol faction that used rule-based method and the random faction that used a probability-based method. Restricted by the data and computing power at the time, the random faction could not reach its full potential, and the symbol faction got the upper hand. When it came to translation, people believed that the process for machine translation was to decipher a code and try to consult a dictionary to gradually translate one word at a time. The results of this translation method were not good, and it was hard to apply. There were some results at the time, including the 1959 University of Pennsylvania Transformation and Discourse Analysis Project (which was the first automatic and complete analysis system for the English language) and Brown University's establishment of an English-language data bank. IBM-701 computer carried out the world's first machine translation experiment, translating a few simple sentences of Russian into English. At this time, Russia, Britain, Japan and other countries all carried out machine translation experiments.

In 1966, the American Academy of Sciences' Automatic Language Processing Association Committee (ALPAC) released a research paper entitled "Language and Machines," the whole thing denying the possibility of machine translation and saying that machine translation could not surmount its current difficulties and would be tough to put into use. This report extinguished the earlier fervor over machine translation, and much related research was forced to stop, with natural language processing falling into a nadir. Many researchers drew a lesson from this painful experience to realize that the difference between two languages did not purely amount to differences in vocabulary, but also included differences in syntax structure. In order to raise the readability of translation texts, they determined to carry out research to strengthen language modelling and semantic analysis. The next milestone was in 1976, when Canada's Montreal University and Canada's federal government's Translation Bureau worked together to develop a machine translation system called TAUM-METEO, which provided a weather forecast service. This system could translate 60,000–300,000 words per hour and would translate 1000–2000 pieces of meteorological data per day, which could be sent out via television or newspapers. After this, the EU and Japan also one after another started to research multiple language machine translation systems, but they did not achieve their expected results.

When the 1990s arrived, natural language processing entered a period of booming growth. As the calculating speed and storage volume of computers increased on a large scale, and with the large-scale creation and accumulation of real texts, as well as the need for information retrieval and collection requirements based on natural language that were sparked by the arrival of the Internet, with webpage search being representative, people's zeal for natural language processing reached an all-time high. In traditional rules-based processing technology, people introduce more data to propel their statistical method, and this allowed natural language processing research to reach new heights. Apart from machine translation, webpage search, speech alternatives, dialogue robots and so on all were contributed as natural language processing.

After 2010, technologies based on big data and shallow or deep learning allowed further optimization of natural language processing. Machine translation results again improved, and specialized intelligent translation products were optimized. Dialogue exchange capabilities were used in service robots and intelligent assistant products. An important milestone at that time was the IBM-developed Watson system taking part in the variety show *Jeopardy!*. In the competition, Watson was not connected to the intern, but could rely on its 4 terabytes disk with a two-million page structured and non-structure information to successfully win against a human contestant, displaying to the world the actual strength of natural language processing technology. As for machine translation, Google's neural network machine translation using the relatively traditional phrase-based machine translation could translate English to Spanish with an error rate dropped to 87 percent while English to Chinese error rate dropped to 58 percent, achieving a very significant improvement.

Planning and Decision-Making Systems

AI planning and decision-making systems' development was to an extent carried out by chess-type games. In the nineteenth century, there first appeared a chess playing machine that defeated almost all human players, including Napoleon and Franklin. But in the end it was discovered that the machine actually had a chess master hidden inside and was just a hoax. The first actual planning and decision-making system came in 1962 after the birth of electronic computing when Arthur Samuel created a checkers process that was altered many times and eventually was able to beat the national champion. At the time the process, although it wasn't intelligent, made a big stir, as it was after all the first time that a machine had beaten a human in a contest of intellect. This also caused people to make optimistic predictions that "machines will within ten years be able to beat the human Chinese chess champion."

But AI faced more difficulties than people imagined. The checkers program later lost to a national champion and could not attain a higher goal. And compared to checkers, international chess is much more complicated. With the computing power of the time, if machines wanted to beat a human chess master via reinforcement calculation then for each move the average calculation time could be measured in units of years. People also realized that only by reducing the computation calculation complexity as much as possible could a machine be superior to a person in each decision. As such, the "trimming method" became used in the valuation function, with elimination of low probability moves being used to optimize the estimated function calculation. With this "trimming method," Northwest University developed the chess program Chess 4.5 which in 1976 for the first time defeated a top-rated human chess champion. Entering into the 1980s, as algorithms continuously were optimized, machine chess programs' judgement capabilities and calculation speed significantly increased, and they could already defeat all top-ranked human chess masters.

When it came to the 1990s, hardware functionality and algorithm capacities had all significantly improved. In 1997 in that famous battle between man and machine, the IBM-developed Deep Blue beat the international chess champion Kasparov, and people realized that in the game of chess it was already very hard for humanity to win against a machine.

In 2016, a dual processing Graphics Processing Unit (GPB)- and Tensor Processing Unit (TPU)-based computation was developed at the hardware level, and at the algorithm level deep neural networks were combined with Monte Carlo decision trees. Humans' last stronghold in strategy games, Go, was also conquered by AlphaGo. Li Shishi was beaten four games to one; top Go champions were beaten 60 times in a row on the Yehu online Go platform; and top-ranked Go player Ke Zhi was beaten three games to zero. Humanity had completely given up to machines in perfect information contests and could find shelter only in imperfect information games such as Texas Hold'em or mahjong. The experience and knowledge accumulated by people from strategy games also have wide-ranging applications in the field of decision-making and planning, including for machine control and unmanned vehicles. Strategy games had fulfilled their historic mission to bring AI to the forefront and begin a new phase of history.

The Third Wave of AI

From the Dartmouth summer AI research conference in 1956, when the concept of AI was first put forward, the development of AI technologies has already gone through 60 years of changes. In these 60 years, the development of AI technologies has not progressed without a hitch and has gone through two waves of development in the 1950s and 1960s, then again in the 1980s, as well as going through a winter in the 1970s and 1980s. Along with the explosive growth of data and big improvements in computing power, as well as the development and maturity of deep learning, we have already reached the third wave of development since the concept of AI was put forward. However, this time is clearly different from the previous two waves. At the same time, the influence of this round of AI development has gone far beyond the academic world, with government, enterprises and non-profits all-embracing AI technologies. AlphaGo beating Lee Sedol also meant that the public had an even greater knowledge and understanding of AI technologies. The third wave of AI development that we are currently in is just the beginning. Today, 60 years since the concept of AI was put forward, the rapid development of AI has pulled forward the curtain onto a new era.

Bibliography

Bengio, Yoshua, Ducharme, Réjean, Vincent, Pascal and Janvin, Christian. "A Neural Probabilistic Language Model." *J. Mach. Learn.* Res. 3 (2003): 1137–1155.

Breiman, Leo. "Random Forests." *Machine Learning* 45 , no. 1 (2001): 5–32.

Breiman, Leo, Friedman, J. H., Olshen, R. A. and Stone, C. J. *Classification and Regression Trees*: Wadsworth, 1984.

Cortes, C. and Vapnik, V. "Support Vector Networks." *Machine Learning* 20 (1995): 273–297.

Freund, Y. and Schapire, R. "A decision-theoretic generalization of on-line learning and an application to boosting." *Journal of Computer and System Sciences* 55, no. 1 (1997): 119–139.

Gantz, John and David Reinsel. *IDC Study: Digital Universe in 2020*, 2012.

Goodfellow, Ian J., Pouget-Abadie, Jean, Mirza, Mehdi, Xu, Bing, Warde-Farley, David, Ozair, Sherjil, Courville, Aaron and Bengio, Yoshua. Generative Adversarial Networks. (2014), cite arxiv:1406.2661 .

Hebb, Donald. *The Organization of Behavior*. New York: Wiley, 1949.

Hinton, G. E. and Salakhutdinov, R. R. "Reducing the Dimensionality of Data with Neural Networks." *Science* 313 (2006): 504–507.

Hochreiter, S., Bengio, Y., Frasconi, P. and Schmidhuber, J. "Gradient flow in recurrent nets: the difficulty of learning long-term dependencies." In *A Field Guide to Dynamical Recurrent Neural Networks* , edited by Kremer and Kolen: IEEE Press, 2001.

Hopfield, J. J. "Neural Networks and physical systems with emergent collective computational abilities." *Proceedings National Academy of Science* 79 (1982): 2554–2558.

Hopfield, John J. "Neurons with Graded Response Have Collective Computational Properties Like Those of Two-State Neurons." *Proceedings of the National Academy of Sciences*, USA 81 (1984): 3088–3092.

Krizhevsky, Alex, Sutskever, Ilya and Hinton, Geoff. "Imagenet classification with deep convolutional neural networks." Paper presented at the meeting of the Advances in Neural Information Processing Systems 25, 2012.

Lecun, Yann and Bengio, Yoshua Convolutional Networks for Images, Speech and Time Series. Arbib, Michael A., eds. *The Handbook of Brain Theory and Neural Networks* , The MIT Press (1995), 255–258 .

Mcculloch, Warren S. and Pitts, Walter H. "A Logical Calculus of the Ideas Immanent in Nervous Activity." *Bulletin of Mathematical Biophysics* 5 (1943): 115–133.

Minsky, Marvin and Papert, Seymour. *Perceptrons: An Introduction to Computational Geometry*. Cambridge, MA, USA: MIT Press, 1969.

Rosenblatt, F. "The perceptron: a probabilistic model for information storage and organization in the brain." *Psychological Review 65* (1958): 386–408.

Rumelhart, David E., Hinton, Geoffrey E. and Williams, Ronald J. "Learning representations by back-propagating errors." *Nature* 323, 6088 (1986): 533–536.

Sutton, R. S. "Learning to predict by the method of temporal differences." *Machine Learning* 3 (1988): 3-9-44.

Sutton, R. S. "Temporal Credit Assignment in Reinforcement Learning." PhD diss., University of Massachusetts, Dept. of Comp. and Inf. Sci., 1984.
Watkins, Christopher John Cornish Hellaby. "Learning from Delayed Rewards." PhD diss., King's College, Cambridge, 1989.
Werbos, Paul. Beyond regression: New tools for prediction and analysis in the behavioral sciences. PhD diss, Harvard University, 1974.
NIPS Workshop: Deep Learning for Speech Recognition and Related Applications, Whistler, BC, Canada, Dec. 2009 (Organizers: Li Deng, Geoff Hinton, D.Yu).

CHAPTER 3

Artificial Intelligence: Today and in the Future

Today, the development of artificial intelligence has already broken through a certain "threshold." Compared with the previous upsurge, this time artificial intelligence has become more dependable, reflected in the performance improvement and efficiency optimization in different vertical fields. The accuracy of computer vision, speech recognition, and natural language processing is no longer stuck at the level of "playing house," or mimicking reality. Applications are no longer just a novel "toy," but are gradually playing a real and important supporting role in the business world.

Speech Processing

A complete speech processing system will include front-end signal processing, intermediate stage speech semantic recognition, and dialogue management (more often involved in natural language processing), as well as later stage speech synthesis. Overall, with the rapid development of speech technology, the previous restricting conditions are continually decreasing: from small vocabulary to large vocabulary to super vocabulary; from a restricted language environment to a flexible environment to an unrestricted environment; from a quiet environment to a normal speaking environment to a noisy environment; from a clear reading environment to a normal speaking environment to an unrestricted conversational environment; from monolingual to multilingual to different languages mixed together—these all put higher demands on speech processing.

Tencent Research Institute et al. (eds.), *Artificial Intelligence*,
https://doi.org/10.1007/978-981-15-6548-9_3

The front-end processing of speech covers several modules. Detecting the voice of the person speaking: effectively detection involves finding the moments when speech starts and finishes, distinguishing between the target voice and background noise. Echo cancellation: when music is playing, in order to effectively perform speech recognition without pausing the music, it is necessary to eliminate interference of music that comes from the speaker. Wake-up word recognition: the means of triggering the start of human-machine communication, just like when starting to talk to other people in everyday life, you will first call out that person's name. Microphone array processing: after locating the sound source, strengthen the signal pointing toward the person speaking and suppress noise signals coming from other directions. Speech enhancement: further enhancing noise coming from the area where the person is speaking and further suppressing the area with ambient noise, effectively reducing the weakening sound levels of distant speech. Aside from handheld devices, which are short-distance interactions, many other scenarios—cars, smart homes—are all long-distance environments. In long-distance environments, the sound is seriously reduced by the time it has reached the microphone, causing a number of obvious problems not found in short-distance environments. This requires front-end processing technology to overcome issues including noise, reverberation, and echoes in order to better pick up sound over long distances. More training data is also required in a long-distance environment in order to continuously optimize the model and to improve efficiency.

Speech recognition needs to undergo multiple processes such as feature extraction, model self-adaptation, acoustic modeling, language modeling, and model decoding. In addition to the long-distance identification issues mentioned above, there are many leading studies that are working to solve the "cocktail party problem," which is an attempt to replicate a human ability to track and recognize one or more sounds among many different people speaking mingled with background noise, such that the noisy environment does not impact normal communication. This ability is found in two scenarios: one is when people focus on a certain sound, such as a friend talking at a cocktail party, even if the surrounding environment is very noisy, or even if the volume of background noise exceeds the volume of your friend, we can still clearly hear what the friend says. The second situation is when people's hearing organs receive a shock, such as when someone suddenly shouts their name in the distance, or if they suddenly hear their native language in an environment where everyone is speaking a non-native language; in these scenarios, even if the voice is far distant and the volume is very small, our ears capture it immediately. A machine lacks this ability; although the current speech technology can show a high degree of precision when recognizing content from an individual, when there are two or more speakers the recognition accuracy will be greatly reduced. Technically speaking, the essence of the problem is how to deal with overlapping speech from multiple people. A simple task is how to separate the signal from a single speaker from the background noise of other speakers, and a more complicated task is to separate out each individual's separate and independent yet simultaneous speech. For such tasks, researchers have proposed some solutions, but they also require accumulated training data and the polishing of training processes in order to gradually make breakthroughs and to finally solve the "cocktail party problem."

Considering that the semantic recognition and dialogue management sectors are more within the scope of natural language processing, the rest is just speech synthesis. The steps of speech synthesis include text analysis, linguistic analysis, sound length estimation, pronunciation parameter estimation, and the like. Speech synthesized using current-day technology is already quite clear and at a good level of clarity and intelligibility, but there is still quite a strong machine "accent." Current research directions include: how to make synthetic speech sound more natural; how to make synthetic speech more expressive; how to allow multiple languages to be synthesized together in a natural and smooth manner. Only a breakthrough in these areas could make the synthesized speech truly sound the same as a human voice.

It can be seen that under certain restrictions, machines really are able to "hear" to a degree. Therefore, in some specific scenarios such as voice search, voice translation, and machine reading, there is ample scope for its use. But it remains the case that, just as for normal human beings, it will take a long time for machines to communicate smoothly and freely exchange speech with other people.

Computer Vision

Computer vision research has moved from problems of easy technical difficulty to difficult ones, has moved from a high level of commercialization to low, and has successively moved from processing to recognition and detection, to analysis and understanding. Image processing refers to processing that does not touch upon high-level semantics, rather it only aims to process the underlying pixels; image recognition and detection includes basic search using speech information; image understanding takes things to a higher level, to include richer, broader, and deeper semantic search. At the processing and recognition levels, machine performance is already satisfactory, but at the level of understanding, there are still many areas of research worth pursuing.

Image processing is based on a large amount of training data (e.g., through the pairing of images either with or without noise). There are several tasks that can be completed end-to-end through the use of deep neural network training, such as denoising, deblurring, super-resolution processing, and filter processing. Applied to video, the main purpose is to filter the video. At present, these technologies are already relatively mature and can be seen everywhere in various photoshop and video processing software.

The process of image recognition and detection includes image preprocessing, image segmentation, feature extraction, and judgment matching. It is also an end-to-end solution based on deep learning, which can be used to deal with classification problems (such as whether the image being recognized in a picture is a cat); location problems (such as identifying the location of the cat in the picture); detection problems (such as identifying which animals are in the picture and where they are); segmentation questions (such as which pixel areas in the picture are cats); and so on. These technologies are also relatively mature, and the applications on the image include face detection and recognition as well as optical character recognition (OCR), while video can be used to recognize movie stars. Deep learning, of course, plays an important role in these tasks. Traditional facial recognition algorithms can only hit an accuracy rate of about 95 percent even if it takes into account the features including color, shape, and texture. With the addition of deep learning, the accuracy rate can reach 99.5 percent and the error rate drops by 4.5 percent which makes the technology widely commercializable with applications in finance and security. In OCR, the traditional recognition method requires a number of preprocessing tasks, such as sharpness judgment, histogram equalization, grayscale, tilt correction, character cutting, and so on, to obtain a clear character image at the appropriate angle, which can then be recognized and turned into text. Deep learning's emergence not only reduces the amount of complicated and time-consuming pre-processing and post-processing work, but it also increases the accuracy rate of character recognition from 60 percent to over 90 percent.

Image understanding is essentially the interaction between images and text and can be used to perform text-based image searches, image description generation, image question and answer (given images and questions, output answers), and more. In the traditional method, the text-based image search returns the corresponding image of text to match the search after searching for the image of text most similar to the inputted text; the image description is generated based on the object that has been recognized from the image; a rule-based template is then used to create the description of the text; and the image question and answer is achieved by obtaining separate digital representations of the image and text respectively, and then separately obtaining answers. With deep learning, you can create an end-to-end model that directly connects the image and the text to achieve better results. Image understanding tasks have not yet achieved very mature results, and commercialization possibilities are currently being explored.

It is notable that computer vision has already reached a stage where it is beginning to be used in entertainment and in machinery. Functions such as automatic image classification, image search, and image description generation can be used as tools to assist human vision. People no longer need to rely solely on information captured by the naked eye then processed, analyzed, and understood by the brain, but can hand over the capture, processing, and analysis to a computer before returning the results to the humans. Looking to the future, computer vision is thought to be able to enter an advanced stage of automatic understanding and to even carry out analysis and decision-making, so as to truly give the machine the ability to "see" and thus be of greater use in applications such as smart homes and unmanned vehicles.

Natural Language Processing

The core link in natural language processing includes the acquisition and expression of acquired knowledge, natural language understanding, and natural language generation. Research directions that have arisen include knowledge mapping, dialogue management, and machine translation, as well as having an illuminating relationship with the aforementioned processing segments that allow many-to-many processing. Since natural language processing requires machines to "understand," which is more complicated than mere "perception," many of these problems still have not been well resolved today.

A knowledge map involves structuring knowledge based upon the semantic layer, and it can be used to answer simple fact-based questions, such as for language knowledge maps (hypernyms, hyponyms, synonyms), common sense knowledge maps ("birds can fly but rabbits cannot fly"), and entity relationship maps ("Andy Lau's wife is Zhu Liqian"). The process of building a knowledge map is actually just the process of acquiring, expressing, and applying knowledge. For example, for a piece of text on the Internet that says, "Andy Lau and his wife Zhu Liqian attended the film festival," we can take out the keywords "Andy Lau," "wife," and "Zhu Liqian," and then get "Andy Lau–wife–Zhu Liqian" to get a ternary representation. Similarly, we can also get a ternary expression like "Andy Lau–height–174cm." Integrating together these ternary representations from different entities in different fields constitutes a knowledge mapping system.

Understanding syntax is the biggest problem in natural language processing. The core issue here is how to find the most suitable way of mapping the many-to-many form and meaning onto the language environment being used. Taking Chinese as an example, there are four difficulties to solve. The first is the elimination of ambiguity, including the ambiguity of words (the word for "diving," for example, can refer to an underwater movement, or it can be used to refer to a "lurker" who does not speak in a public forum), the ambiguity of a phrase (jinkou caidian can mean an "imported color TV," a noun, but it can also mean the action of importing a color TV) and sentence ambiguity (the same sentence can mean both "his father is a surgeon" and "his father is undergoing surgery"). Second is the contextuality, including referencing (e.g., in the sentence, "Joe bullied Jim, so I criticized him," one needs to rely on the context to know that the criticism is of the naughty Jim), the omission of recurring phrases (in the sentence "Old Wang's son is good at learning, better than Old Zhang's" we know that it means "better than Old Zhang's son"). Then there is recognition of intent, including the intent to identify nouns versus words that are simple descriptions ("sunny" could refer to the weather or it could refer to a Jay Chou song); identifying the intention of small talk versus a question ("It's raining today" is small talk, and "It's raining today?" is a query about the weather); identifying tended explicit or hidden meanings ("I want to buy a mobile phone" and "I've used this mobile phone for too long" both convey the intention of the user to buy a new mobile phone). Finally, there is the recognition of emotions, including recognizing both explicit and implicit emotions ("I am not happy" and "I didn't do well in the test" both convey feelings of being in a bad mood), the recognition of emotions that are based on a priori common sense ("a long time" in the sense a battery is positive, while a "long time" in the context of waiting for a flight is negative). In view of the above difficulties, one possible solution to semantic understanding is to use our knowledge of how language works to constrain the difficulties of many-to-many mapping and to supplement the knowledge of machines through knowledge maps that we provide. However, even if the difficulties in understanding semantics can be overcome, it will still be a long way from making a machine not appear to be quite stupid.

It is also necessary to make breakthroughs in dialogue management.

At present, dialogue management, which is used in conversations systems such as chatbots, chiefly involves three scenarios, which can be classified according to whether they involve general or professional knowledge, from a casual conversational, question and answer, or task-driven dialogue. Chatting is an open-ended conversation, involving emotional connections and an individual's personality. For example, in an exchange like, "The weather is really good today"; "Yes, do you want to go out for a walk?" the question is how to either arouse interest or reduce disinterest from a subtle answer and thus extend the dialogue time and improve stickiness. Question and answer is dialogue based on a Q&A model and information retrieval, usually involving only a single round, such as "Who is Andy Lau's wife?" and "Andy Lau's wife is Zhu Liqian, born on April 6, 1966, in Penang, Malaysia, etc." Q&A not only requires a more complete knowledge graph, but also needs to be able to infer answers when a direct reply is not given. Task-driven dialogue involves slot-filling (to return all information about an object when prompted, such as supplying the age, birthplace, occupation, and so on of a particular person) and intelligent decision making. This kind of dialogue usually involves a number of rounds, for example, "play a song that's good to listen to while running"; "I recommend Yu Quan's 'Run' "; "I want to listen to an English song"; "I recommend you Eminem's 'Not afraid.'" Simple task-driven dialogue technology is relatively mature, and the future direction of the capture is how to build a common sphere of dialogue management that doesn't relying on slot definitions to be supplied manually.

Historically, the typical application of natural language generation (creating language from structured data) has always been machine translation.

The traditional method is called Phrase-Based Machine Translation (PBMT): it first breaks a complete sentence into several phrases, then it translates the phrases separately, sorts them according to grammatical rules and finally restores them into a coherent sentence of translated text. The whole process does not seem complicated, but it involves multiple natural language processing algorithms. For Chinese, it would include Chinese word segmentation, part-of-speech tagging (marking up words as verbs, nouns, adjectives, etc.) and working out the syntactic structure, among others. Errors in any of these stages would be passed along and affect the final result. Deep learning relies on a large amount of training data to directly establish a mapping relationship between the source language and the target language via an end-to-end learning process that skips the intermediary steps of complex feature selection and manual tuning. Out of such an idea, the "encoder-decoder" neural machine translation structure was proposed in the early 1990s and has been continuously improved. ("Encoder/decoder" is a hardware tool that can convert knowledge into code and then back again.) Attention mechanisms (a way of allowing machines to focus on select information, rather than being overwhelmed by the whole) have also been introduced to significantly improve system performance. Later, the Google team replaced the previous SMT (Statistical Machine Translation) with a new machine translation system called GNMT (Google Neural Machine Translation), which is much smoother and has a significantly reduced error rate. Although there are still many problems to be resolved, such as the translation of uncommon words, missing words, or repeat translations, it is undeniable that neural machine translation has led to great breakthroughs in performance. In future, application prospects for overseas travel, business meetings, cross-border exchanges, and other scenarios will be considerable.

With the rise of the Internet, the degree of electronic information available has also increased. Massive volumes of data are not only a fuel for training natural language processes; they also provide a grand arena for development to take place. Search engines, chat bots, and machine translation—or even robots that do college entrance examinations and intelligent office secretaries—are beginning to play an increasingly important role in people's everyday lives.

Machine Learning

When considering the various levels of artificial intelligence, machine learning is a lower-level concept compared to the technical layers of computer vision, natural language processing, and speech processing. In recent years, progress at the technical layers has flourished, and machine learning at the algorithm level has also produced several important avenues of research.

The first avenue is the widespread application of machine learning in particular verticals or unique domains. In light of the many limitations of machine learning and its inability to be universally applied across domains, application in a relatively narrow domain has become a good point of entry for machine learning. Because in a limited domain, the problem space is sufficiently small, allowing better results for the model being used. The second application is to use training data from a specific environment, making it easier to accumulate content, which in turn makes the training model more efficient and targeted. The third application is when people have a set and specific expectation for the machine, and that expected value is not high. In these three conditions, the machine can show enough intelligence in this limited field and allow a relatively good end user experience. Therefore, in the unique domains of finance, law, and medicine, we have seen some mature applications that have been commercialized to a degree. In the domain of repetitive manual labor, it is likely that a large proportion of labor will be supplanted by artificial intelligence in the future.

The second avenue ranges from attempts to solve simple convex optimization problems through to attempts to solve non-convex optimization problems. A convex optimization problem is to represent all considered factors as a set of functions, and then choose an optimal solution from them. A positive feature of convex optimization problems is that the local optimum is also the global optimal. At present, most of the problems in machine learning can be transformed into or approximated as a convex optimization problem by adding certain constraints. Although the optimal value of any optimization problem can be found by traversing all the points on the function, the amount of calculation required to solve the problem this way is often huge. Especially when there are many feature dimensions, dimensional disasters occur (the number of features exceeds the upper limit of the number of features that can exist in the known sample number, which leads to the degradation of the performance of the classifier). The characteristics of convex optimization make it possible to find the direction of the decline by the gradient descent method, and the local optimal solution found will be the global optimal solution. However, in real life, there are not many problems that really conform to convex optimization. At present, the focus on convex optimization problems is simple because such problems are easier to solve, just like when people lose keys in the street at night, they will look under the light first. Therefore, in other words, people still lack effective algorithms for non-convex optimization problems, which is also the direction of people's efforts.

Another problem is the evolution of supervised learning to unsupervised learning and reinforcement learning. At present, most AI applications use a set of already labeled training data in a supervised learning environment to adjust the parameters of the classifier to reach the demanded function. But in real life, supervised learning is insufficient to reach the level of "intelligence." The human learning process, for comparison, is mostly based on interactions with objects, which, via human experience and comprehension, allow people to understand those objects, and then to make use of them in future life. A limitation of machines is their lack of this "common sense." Yann LeCun, Facebook's chief AI scientist, used the metaphor of a cake to describe his understanding of the relationship between supervised learning, unsupervised learning, and reinforcement learning: if machine learning is regarded as a cake, then (pure) reinforcement learning is only a cherry on top of the cake, as the sample size required is only a few bits of data; supervised learning is the icing of the cake, requiring a sample size of 10 to 10,000 bits; and unsupervised learning is the main body of the cake, which requires a sample size of millions of bits and has formidable predictive power. But he has also stressed that the "cherries" are an indispensable ingredient, meaning that reinforcement learning and unsupervised learning are complementary and cannot work without each other. Recent research in the field of unsupervised learning focuses on the "Generative Adversarial Networks" (GANs), which involve pitting two networks—one generator and one discriminator—against each other. The generator randomly selects true data and interference noise from the training set, in order to generate new training samples which attempt to "fool" the discriminator, which is attempting to judge the authenticity of the generative data set by comparing it to true data. Through this zero-sum competitive process, the generator and the discriminator automatically optimize their joint predictive ability and can create accurate predictive models. Since Ian Goodfellow's paper popularized GANs in 2014, the idea has taken top AI conferences by storm and was described by Yann LeCun as "the most interesting idea in the last 10 years in machine learning." Reinforcement learning, on the other hand, is closer in its origins to nature's biological learning process: if you think of yourself as an agent in a learning environment, you will want to constantly explore in order to discover new probabilities, and you will want to try to reach the peak action under the existing conditions, known as exploitation. The right decision will, sooner or later, bring about a positive reward, while the wrong decision will bring about a negative reward. Over time, you will begin to develop a thorough understanding of the problem or an optimal policy. An important area of research for reinforcement learning is to establish a simulation environment that allows machines to effectively interact with the real world, so as to continuously learn, to simulate and adopt various actions, and then to take various feedback on board so as to continue training the model.

Ubiquitous Artificial Intelligence Algorithms

With recent successes of deep learning in the fields of computer vision, speech recognition, and natural language processing, the applications of AI algorithms have matured and begun to seep into all aspects of our lives, Intelligent assistants in our smartphones, intelligent recommendations on websites, intelligent investment, and intelligent security systems all rely on AI technology based on machine learning algorithms. AI algorithms are in people's mobile phones and laptops; in the servers of government agencies, companies, and social enterprises; and in shared or private cloud computing. Although we may not always be aware of the existence of AI algorithms, they have become part of our lives. The maturity of artificial intelligence technology in various fields means that in the future, AI technology will increasingly touch upon all aspects of life and will combine with traditional computing models to enhance productivity. As AI is integrated into our lives, it will increasingly disrupt how we live.

The Future of Artificial Intelligence

With the recent rapid advancement of multiple underpinning technologies, AI has finally been ushered in its golden era. Looking back at the ups and downs of artificial intelligence in the past 60 years, we can draw a number of lessons from history. Firstly, infrastructure plays a huge role in pushing the field forward. Artificial intelligence research has repeatedly been left cold due to the limitations of data, computing power, and algorithms, while breakthroughs come from the infrastructure layer facilitating industry application. Secondly, AI in games plays an important role in the advancement of AI, because the game involves person-machine confrontation, which can help people intuitively understand AI perception and action and thus promote new developments. Finally, we must be soberly aware that, although artificial intelligence has achieved results that rival or even surpass humanity on many tasks, there are still a number of obvious bottlenecks. For example, in computer vision, there is the problem of dealing with natural conditions, such as light and occlusion; in subject recognition and judgment problems, there is the problem of how to focus on a target object in a complex picture; in speech recognition technology, there is the problem of dealing with noise outside of specific environments, such as the car or home; and so on. Overall, we see that current existing AI technology is dependent on a large amount of high-quality training data, and furthermore that its ability to deal with the long-tail problem is poor. It is also not very versatile and is dependent on discrete and specified application scenarios.

In the future, people will not only use artificial intelligence to solve simple and specific tasks in a narrow and designated field, but will want it to be like real human beings, which can solve problems from different fields and of different kinds simultaneously to reach judgments and make decisions—which is to say that it will be so-called general-purpose artificial intelligence. Specifically, the machine needs to understand the world via both perceptual and cognitive learning; on the other hand, it can simulate real-world scenarios through reinforcement learning. The former allows the machine to perceive information and transform the perceived information into abstract knowledge through attention, memory, and understanding. The latter is a way to obtain and then optimize knowledge via interactions between the machine and the simulated environment that has been created. People hope that through the intersection, integration, and optimization of algorithms and academic study, the problems of allowing AI to have creative abilities, general-purpose use, and an understanding of objects in the world can be solved.

Let us return to the idea of layers within AI. Looking to the future, the lowest infrastructure layer will provide the data from the Internet and the Internet of things to modern AI environments. The algorithm layer will use deep learning and reinforcement learning to provide the core model of modern AI with an engine powered by cloud computing. On top of this, whether its computer perception, natural language processing, or speech technologies—game AI or robots—all will be based on the same data and models with applications in different scenarios. In this process a number of pressing questions remain. Resolving these questions will take us, one step at a time, on the road toward artificial general intelligence.

The first problem is to go from big data to small data. The training process for deep learning requires a vast amount of data, which has to have been tagged by people. For example, autonomous driving requires a huge number of pictures of streets where people have labeled cars, people, and buildings. For speech recognition, the research needs text to speech broadcasts and speech to text writing examples. Machine translation requires dual-language sentence pairs. To learn how to play chess, AI requires the play records of human champions. But labeling large amounts of data is time-consuming and expensive work, especially in a long-tail environment, when even the collection of foundational data is a problem. As such, one area of research is how to allow training to take place in an environment that lacks data, and to learn from unlabeled data or to automatically simulate the data needed for training, as with the currently very hot GANs model of data creation.

Another problem is how to go from a big model to a small model. Currently, deep learning models are all very large and frequently require hundreds of megabytes, with some larger ones requiring thousands of megabytes or even tens of gigabytes. Although this is fine for models run on a desktop computer, if they need to be easily transportable, then it becomes very difficult. This has prevented mobile apps for speech input, speech translation, and image filtering from reaching high levels of efficiency. This block of research focuses on how to reduce the scale of the models, whether via directly compressing or by finding ingenious models, so as to close the gap between mobile terminals with low-consumption computation and cloud computation, to make small models that are as efficient as larger ones.

Finally, there is the problem of going from perceptual knowledge to understanding and decision-making. In the parts of perception and cognition, such as vision and hearing, machines have been able to reach good outcomes under certain conditions. Of course, these tasks are not difficult at all. The value of machines is that they can do it faster, more accurately, and at a lower cost than people. But these tasks are basically static, that is, given particular inputs, the output is certain. In some dynamic tasks—how to win a game of Go, how to drive from one intersection to another, how to invest in a stock so as to make money—this type of decision-making problem in an incomplete information environment needs continuous interaction with the environment to gather feedback and optimize strategies. These are some of the strengths of reinforcement learning. And simulated environments (simulators) are also an important area of research and are the fertile soil from which reinforcement learning can grow.

In March 2016, when AlphaGo defeated the world champion Lee Sedol, we were all witnessing history. AlphaGo's victory marks the beginning of a new era. Sixty years after the concept of artificial intelligence was proposed, it has now truly entered a new era. In this wave, artificial intelligence technology continues to develop at great speed and will eventually profoundly change multiple aspects of life for everyone. The ultimate goal of developing artificial intelligence is not to replace human intelligence, but to enhance human intelligence through artificial intelligence. Artificial intelligence can complement human intelligence and help humans deal with many tasks that need to be handled, but that is not easy for humans. As such, people can be freed from strenuous repetitive work and instead focus on creative work. With the help of artificial intelligence, people will enter a phase of accelerated knowledge accumulation that will eventually bring about progress in all kinds of different places. Artificial intelligence has brought many surprises and expectations to people on this road of development. As long as we can make good use of artificial intelligence, we can have faith that in the near future, it will certainly realize more things that we thought were impossible and lead humanity into a new era of infinite possibilities.

PART II

Industry: The Complete Picture of the Development of AI

With AlphaGo defeating the best Go players among humans, artificial intelligence has become one of the hottest words of 2017. Undoubtedly, the development of artificial intelligence is inseparable from the strong support of national strategies and policies, and it is inseparable from the development of machine learning algorithms, the improvement of computing power, the continuous opening up of data and the deepening of applications. From the perspective of commercial maturity, transportation, medical care, finance, and entertainment may be the first areas where artificial intelligence will land. Applications such as autonomous driving, intelligent robots, virtual reality, and augmented reality combine many artificial intelligence technologies such as image recognition, speech recognition, and intelligent interaction. They have received high levels of attention from industry and the rest of China, and this will be the focus of this section.

CHAPTER 4

An Overview of the Artificial Intelligence Industry

The battle between industry giants is the main force propelling the technical race in AI. Because core technology and resources of the AI industry are concentrated in the hands of major enterprises—and these resources and structure of the major players in the industry cannot be matched by startups—the big technology corporations lead the field for AI.

At present, the five tech giants—Apple, Google, Microsoft, Amazon, and Facebook—are all investing ever more toward seizing a section of the artificial intelligence market, and they have even started to transform themselves into artificial intelligence–driven companies. The Chinese domestic Internet leaders "BAT"—Baidu, Alibaba, and Tencent—also regard artificial intelligence as a key part of their strategy and have proactively taken steps in the field of artificial intelligence, drawing upon their own industry advantages.

The broad structure of technical competition between the United States and China has already begun to emerge from the dual promotion of new technologies by governments and industry. This chapter analyzes the development of artificial intelligence–related industry in China and the United States.

Tencent Research Institute et al. (eds.), *Artificial Intelligence*,
https://doi.org/10.1007/978-981-15-6548-9_4

The United States Is Leading the World in AI Companies

The United States, China, and other developed countries are leading the global development of artificial intelligence. As of June 2017, the total number of artificial intelligence companies worldwide reached 2542, of which 1078 were in the United States, accounting for 42 percent of the total. China comes second, with 592, accounting for 23 percent, leaving a difference of 486 companies between China and the United States. The remaining 872 companies are based in Sweden, Singapore, Japan, the United Kingdom, Australia, Israel, India, and a handful of other countries.

When looking at historical statistics for enterprise growth, we can see that the development of American artificial intelligence enterprises began five years earlier than China. AI development in the United States first sprouted in 1991, then entered an early development period in 1998, before beginning rapid growth in 2005. After 2013, it has stabilized. The Chinese AI company was born in 1996 and entered the development period in 2003. After reaching its peak in 2015, industry development has stabilized.

The United States has a complete industry, while China has only made some partial breakthroughs.

The AI industry in the United States is the clear overall leader. It has accumulated strong technological advantages in all layers, from foundational technology up to application, and especially in core capabilities like algorithms, chips, and data.

At all levels and in all fields, the United States is leading compared to China when measured by the number of companies.

China has 14 companies that work on the foundational layer of manufacturing processors and chips, only 42 percent of the 33 US companies doing similar work.

For the technical layer, which includes natural language processing, computer vision, and image recognition, China has 273 companies compared to the United States' 586.

At the level of application—machine learning application, intelligent drones, intelligent robots, autonomous and assisted driving, and speech recognition—China has 304, while the United States has 488.

In America, the pool of talent is complete, but in China it is uneven.

Competition in the AI industry is, when you get down to it, a competition for talent and for a store of knowledge. Only by investing more in researchers and continually strengthening basic research can companies hope to gain more intelligent technology.

America's researchers pay more attention to basic research, the country's artificial intelligence personnel training system is solid, and their research talents have significant advantages. Specifically, in the key links of basic discipline construction, patent and paper publication, high-end research and development talents, venture capital, and leading enterprises, the United States has formed a pattern that can lead the world for a long time.

The total amount of industrial talent in the United States is about twice that of China. There are about 78,000 employees in 1078 artificial intelligence companies in the United States, and about 39,000 employees in 592 companies in China. The number of basic talents in the United States is 13.8 times that of China. The number of American teams has completely suppressed China in the four hotspots of processor/chip, machine learning application, natural language processing, and intelligent drone.

In the field of research, in recent years, China's papers and patents in the field of artificial intelligence have maintained rapid growth and have entered the first echelon. In comparison, China's artificial intelligence needs continuous investment in R&D costs and scaling of R&D personnel as well as increasing talent training in basic disciplines, especially in the areas of algorithms and computing power.

The United States has invested heavily in capital, and China has been catching up in recent years. Startups tend to be the prey of the giants. For example, if the AI industry is a huge machine, then the emerging startups are mostly a part of the machine. This is because emerging startups have only one or several technical advantages, so it is difficult to become a dominant global application, but they help to improve the ecosystems of the giants and thus it is ultimately difficult to escape the fate of being acquired.

Since the first venture capital in artificial intelligence in the United States in 1999, global AI has accelerated. In 18 years, the total amount of venture capital invested in artificial intelligence has reached $191.4 billion. Up to now, American companies have received $97.8 billion of this, leading Chinese companies by 54.01 percent and accounting for 51.10 percent of global financing. China is second only to the United States with investments reaching $63.5 billion, accounting for 33.18 percent of the global total; other countries account for 15.72 percent. There have been more $100 million+ investments in China than that of the United States

(22 vs. 11), but the total value of these large-scale investments is higher in the United States ($41.73 billion vs. $35.35 billion).

The trend of giant companies investing in talent and technology through investment and mergers and acquisitions of artificial intelligence is becoming more and more obvious. Sino-US mergers and acquisitions have increased intensively in the past two years. CB Insights' research report shows that Google has acquired 11 artificial intelligence startups since 2012, the largest among all technology giants, with Apple, Facebook, and Intel ranking second, third, and fourth respectively. The targets are concentrated in the fields of computer vision, image recognition, and semantic recognition. Google acquired DeepMind, a deep learning algorithm company, for $400 million in 2014. DeepMind's AlphaGo adds a touch of color to Google's artificial intelligence.

China and the United States Have Their Own Advantages in the Main AI Hotspots

Deep learning has led the development of this round of AI. The reason is that computing power and data have made major breakthroughs in the past decade. At present, the artificial intelligence industry has emerged in nine hotspots, namely chip, natural language processing, speech recognition, machine learning applications, computer vision and image, technology platform, intelligent drone, intelligent robot, and automatic driving.

The top three areas in American AI startups are natural language processing, machine learning applications, and computer vision and imaging. The top three areas in China's AI startups are computer vision and imaging, intelligent robotics, and natural language processing.

US Leading Industry Giants Have First-Mover Advantage

The giants accelerate the development of key technologies by recruiting AI high-end talents and setting up laboratories. At the same time, they continue to acquire emerging AI startups, compete for talents and technologies, and build an ecosystem through open source technology platforms.

What Is the Future of China's AI Industry?

In the IT era, the Wintel Alliance of Windows and Intel became the standard everywhere; in the Internet era, Google and Amazon sprung up and dominated the world; in the mobile era, Apple and Google are again leading the world. Now, artificial intelligence is slowly bringing in a new chapter.

Similar to the Internet, China will become the largest market for AI applications, with a wealth of application scenarios, with the world's largest number of users and active data production entities. We need to further increase the construction of basic disciplines and personnel training so that Chinese AI can have a chance to go further.

The improvement of national strength comes from the innovation of technology enterprises. The United States is in a leading position with absolute strength, and a group of Chinese startups are also poised for growth. In the future, the AI era will inevitably produce global companies like Intel, Microsoft, Google, and Apple. We believe that Chinese companies have the opportunity to ride the wave of the artificial intelligence era and have a place in the AI field.

AI is a competitive landscape, the world is undecided, and both opportunities and challenges lie ahead. Let us keep a cool head and witness this great era.

CHAPTER 5

Autonomous Driving

In crowded cities, many people feel that driving is a pain: people with bad sense of direction, working professionals who need to attend lots of social engagements, the elderly who are slow to respond all contribute to this feeling. For them, they cannot call a cab during rush hour, the subway is too crowded, bicycles are unsafe, and traffic has become a shared headache for the modern city. With self-driving cars, these troubles can be easily solved. Maybe in the near future, we can call a driverless car to pick us up and deliver us safely to our destination.

The advent of autonomous driving technology could be world-changing. In the automotive industry, for example, autonomous vehicles may shift from an industry targeting individual buyers where car companies sell private vehicles, to one that provides vehicle services for things like entertainment. In the information and technology industry, self-driving cars will be connected to each other through communication technologies, like 5G, and autonomous driving services will also be on offer in mobile communication business hall. In finance, given that autonomous vehicles will rarely ever have accidents, the definition of car insurance as well as the industry's funding and structure will undergo tremendous changes. The technologies also raise new questions for government traffic regulation, such as, since the driver is no longer a human, can the autonomous driving license be canceled? For the industries involved, these changes are no longer a distant dream. Multinational tech giants such as Google and Apple are already moving in this direction, and the governments of the United States, Germany, Japan, and China are also actively

Tencent Research Institute et al. (eds.), *Artificial Intelligence*,
https://doi.org/10.1007/978-981-15-6548-9_5

working on autonomous driving projects, in the hope that they can take the lead in developing and deploying these technologies.

Overall, autonomous driving has grown out of deep integration between the automotive industry and artificial intelligence, the Internet of Things, high-performance computing, and other forms of next-generation information technology. It is now a core pillar of development for transportation and the automotive industry globally, as the two spaces integrate intelligence and connectivity.

The Elements of Autonomous Driving

Autonomous vehicles can be understood as four-wheeled robots. They use sensors, cameras, radar-aware environments, GPS, and high-precision maps to determine their location, as well as receiving traffic information from cloud databases. These various data are collected and processed in order to send commands to the control system which in turn carries out operations like accelerating, braking, changing lanes, or following the car in front.

Levels of Autonomous Driving Technology

Autonomous driving technology is commonly divided into multiple levels depending on its degree of autonomy. The industry most often uses the classification standards developed by the US Society of Automotive Engineers (SAE) or those of the US National Highway Traffic Safety Administration (NHTSA). According to the SAE standards, self-driving cars can be classified into six levels according to the degree of intelligence and automation: no driving automation (L0), driver assistance (L1), partial driving automation (L2), conditional driving automation (L3), high driving automation (L4) and full driving automation (L5).[1].

Two Roads Toward Autonomous Driving

Broadly speaking, there are two different approaches to developing autonomous driving technology. The first is via gradual evolution, which is to gradually add degrees of autonomous driving functions to cars that are

[1]The US Department of Transportation adopted the SAE autonomous driving levels taxonomy.

already on the road today. This approach has been adopted by companies like Tesla, BMW, Audi, and Ford. The method mainly uses sensors, vehicle-to-vehicle communication (V2V), and vehicle-to-cloud communication to carry out road condition analysis. The other route is that of complete transformation, which aims to launch a fully autonomous vehicle from the outset. This is the approach taken by Google, where they are testing such vehicles on structured roads, such as highways or known city grids. These fully autonomous vehicles rely mainly on vehicle-mounted distance-measuring reflected light sensors, known as Lightlaser detection and ranging or lidars, as well as computers and control systems. In terms of application, the first route is more suitable for testing in structured road environments, while the second can be used for military or specialist applications in addition to in structured road environments.[2]

The Software and Hardware Involved in Autonomous Driving

Sensors

Sensors are the eyes of autonomous vehicles. Using sensors, autonomous vehicles can identify roads, other vehicles, pedestrian obstacles, and basic traffic infrastructure, in order to ensure that the vehicle has perceived its surrounding environment with a minimal degree of testing and verification. Depending on the different technical route toward autonomous driving being taken, the choice of sensors can shift between a combination of lidars, traditional radars, and cameras.

Lidar is currently the most commonly used type of sensor. The autonomous driving technology of companies like Google, Baidu, and Uber relies on lidars. The device is most often installed on the roof of a car, where it can detect the surrounding environment with laser pulses that allow inbuilt software to draw 3D images of the environment, in order to provide sufficient information for the vehicle to drive autonomously. Lidars have a high degree of accuracy and speed when identifying objects, but they are expensive, with an average price of $80,000, which makes it difficult to use this technology in mass production cars.

[2] Structured roads refer to roads with fairly standard verges, even surfaces, and clear lane and other markings, for example, highways, urban trunks, and so on.

Traditional radars and cameras are alternative sensor options. Due to the high price of lidars, the companies that are taking the more practical step-by-step development route have turned to the use of traditional radars and cameras instead, using the vehicle's software and connectivity capabilities to compensate. For example, Tesla uses radar and uni-directional cameras. The principle underlying the hardware is similar to that of existing adaptive cruise control systems. The cameras and the front radar cover a 360-degree view of the car's environment and are used to provide three-dimensional information of surrounding object, so as to ensure that the vehicle does not collide with other vehicles. This solution is low-cost and easy to mass-produce, but it relies heavily on the cameras for recognition: these uni-directional cameras need to establish and maintain a large database of sample features. If this feature data lacks the ability to recognize a target object, it will result in the system failing to recognize and adopt the required range, which can easily lead to accidents. One way to improve problems of distance and depth is to use binocular cameras, which can assess depth in a similar manner to the human eye and directly measure the environment in front of the vehicle. However, the amount of calculation required for this is much greater than for uni-directional cameras, and the calculation capability needs to be improved in order for this to be an option.

Automated vehicles can navigate only if they are able to accurately identify their own location, so the importance of maps is self-evident. The basis for autonomous vehicles to understand their environment is data about lanes, distances, and obstacles on the road, making more accurate location information increasingly important. As autonomous driving continues to evolve, safe decision-making requires centimeter precision. If the sensors provide an intuitive sense of the vehicle's environment, high-precision maps can use the precise position of the vehicle within a dynamic three-dimensional traffic environment to work out its exact location.

There are two main ways to select map routes: one is by using high-definition (HD) maps. Such maps are often included in manufacturer programs that use lidars to create a 360-degree awareness of the vehicle's environment. The second is to use a map with features. This method is often combined with radar and camera solutions with maps that include information about lane markings, routes, and road markers. Although this approach provides lower map accuracy, the ability to highlight road features makes vehicles' system processing and updates more convenient.

Map providers also need to continuously collect and update sensor packages and environment data for this system to be efficient.

There are also two main options when determining the positioning of vehicles. The first is to use high-definition maps, which allow vehicle-mounted sensors, including GPS, to compare the environment perceived by the autonomous vehicle and that shown on the high-definition map, thereby accurately identifying the location of the vehicle, as well as its lane and direction of travel. Within this methodology are technologies like vehicle-to-everything communication, or V2X, where information is passed between the vehicles and any object that might affect it.[3] The second option is through the use of GPS positioning. This approach works out the position of the vehicle mainly through GPS positioning and then uses devices such as the vehicle's camera to improve the given positioning information, with frame-by-frame comparison of the two data sources being used to reduce the range of error from the GPS signal. Both positioning methods rely on navigation systems and mapping data. The first provides a more accurate location information, but the second method is easier to deploy and does not require support from high-precision maps. From an engineering perspective, the second method is more suitable for rural or sparsely populated areas, given the lower accuracy of the vehicle's positioning information.

Decision-Making

Autonomous vehicle engineers currently use a range of methods to make autonomous driving decisions. The first is to use neural networks, which primarily identify specific scenarios and make appropriate decisions; however, the complexity of these networks often makes it difficult to understand the foundational reason or logic of a particular decision. The second method is to use a rule-based decision-making system, mostly in the format of an "if-then" decision-making system where choices are made according to specified rules. The third method is hybrid decision-making, which includes both of the above two decision-making methods. This is most often achieved by a centralized neural network that is connected to

[3]V2X refers to the technology by which a vehicle interacts with the surrounding traffic control system. X could be a vehicle, a traffic light, or other transportation infrastructure, or a cloud database. The ultimate goal is to help the self-driving vehicle master real-time driving information and traffic information.

processing by an individual, where a set of "if-then" rules improve the chosen route.

For all of the above methods, the algorithm is critical to supporting the decisions of autonomous driving technologies. At present, mainstream autonomous driving companies use machine learning and artificial intelligence algorithms to make decisions. Massive volumes of data are the foundation of machine learning and artificial intelligence algorithms. This data is obtained from the sensors, V2X equipment, and high-precision map information, as well as from data collected on driving behavior, driving experience, driving rules, individual cases, and the surrounding environment. This data allows the algorithms to continuously optimize and ultimately to identify and plan routes that can be used to drive.

Trends in Autonomous Driving

Overall, the United States and Germany lead the way in developing autonomous driving, with which Japan, South Korea, and China are catching up. Below are specific trends.

Accelerate road testing, as well as relevant laws and regulations, with a goal of commercializing as soon as possible

Countries have taken 2020 as an important turning point by which they hope to have begun the full deployment of autonomous vehicles. The United States is already actively drafting autonomous driving legislation at the federal and state levels. On July 27, 2017, the US federal government made a major breakthrough in the legislation on autonomous driving, when the House of Representatives unanimously passed the SELF DRIVE Act, the first attempt globally to regulate the production, testing, and public release of autonomous vehicles.[4] If it is approved by the US President, the bill could become law and be officially implemented.[5] In

[4] This bill was mainly to amend the 49th Articles of Transportation in the United States Code. The key points are the contents of Chapters 4 (safety standards), 5 (cyber security requirements), and 6 (legal exemptions for self-driving cars).

[5] The procedure for the creation of any law in the United States is: First, a bill is proposed by a member of the US Congress. When the bill is passed by Congress, it will be submitted to the President of the United States for approval. Once the bill is approved by the President, it becomes law. When a law is passed, the House of Representatives publishes the content of the law in the United States Code.

the meantime, there has been a bevy of state legislation governing autonomous driving, with 20 states having enacted 40 bills and administrative orders regarding the technology as of August 2017.[6]

In 2015, the German government approved autonomous vehicle testing on the A9 motorway that connects Munich and Berlin. In April 2016, the German Ministry of Communications drafted a bill to extend the definition of "driver" to include systems with total autonomous control of a vehicle. In May 2017, the German Federal Senate voted to pass the first autonomous driving law, allowing the technology to replace human driving under certain conditions.

As for China, the Ministry of Industry and Information Technology in 2016 launched a number of demonstration projects, including a Shanghai trial for "intelligent networked vehicles," as well as further trials of "the application of smart transportation and intelligent vehicles using broadband mobile internet" in the provinces and municipalities of Zhejiang, Beijing, Hebei, Chongqing, Jilin, and Hubei, among others, with the aim of promoting road tests. Beijing has issued a five-year action plan for the demonstration of smart cars and smart transportation applications. It will complete the transformation of the smart road network of all arterial roads within the Beijing Development Zone by the end of 2020, and deploy 1000 fully autonomous vehicles in phases. In November 2016, Jiangsu signed an agreement with the Ministry of Industry and Information Technology and the Ministry of Public Security to jointly build a "comprehensive national test base for intelligent transportation."

Toward connected cars: Pushing for systematic research and unified telecommunications standards

As it currently stands, most companies have adopted connected cars as their chosen route of further development, which will require an increase in chip processing power, the development of more aware autonomous driving systems, and the introduction of unified vehicle telecommunication standards.

When it comes to research and development, the German Bosch Group and American technology company NVIDIA are working together to

[6] State-level legislation mainly covers commercial deployment, vehicle network security, and ten other aspects. http://www.ncsl.org/research/transportation/autonomous-vehicles-legislative-database.aspx.

develop an artificial intelligence autonomous driving system. NVIDIA provides deep learning software and hardware, at the same time as Bosch AI will be based on NVIDIA Drive PX technology and NVIDIA has also developed a super chip Xavier. These technologies will be combined to reach level 4 autonomous driving. IBM, in March 2018, announced that its scientists had been granted a patent for a machine learning system that helps prevent accidents by allowing a vehicle's control to dynamically shift between autonomous driving processes and that of a human driver in the event of a potential emergency.

Telecommunication technologies, including wireless connections such as LTE-V and 5G, are now key for creating the required telecommunication standards for autonomous vehicles across the industry. These new technologies will provide high-speed, low-latency network support for autonomous driving. The inclusion of LTE-V2X has been part of 4.5G rollouts both in China and elsewhere, with Chinese entities such as China Datong Corp, Huawei, and the China Information and Communication Research Institute pushing for progress in standardizing vehicle-to-vehicle and vehicle-to-infrastructure practices. LTE-V2X technology is also gradually evolving toward 5GV2X as demand for autonomous driving develops. Dedicated V2X communications facilitated by 5G can extend the sensory range of autonomous vehicles beyond the normally working boundary of the sensors, so as to bring about autonomous driving and also safer industrial applications of the technology. This development will help complete the shift of vehicles from being merely a tool for travel toward being an information and entertainment platform, allowing a greater number of business applications. Currently, the 5G Automotive Association (5GAA) and the European Automotive and Telecoms Alliance (EATA) have signed a memorandum of understanding to jointly advance the cellular V2X industry, by working on standardization, spectrum, and use cases related to autonomous driving. In China, partnerships between China Mobile and BAIC, GM and Audi, as well as Huawei's cooperation agreements with BMW and Audi, promote 5G development and deployment. Furthermore, a guiding document on the building of a system from intelligent networked vehicle standards released in 2017 by China's Ministry of Industry and Information Technology is pushing for progress in forming a system of vehicle networking standards, which are crucial for the development of intelligent networked vehicles.

By adopting innovative approaches to autonomous driving, Internet companies lead the way

Internet companies come into being with innovation inbuilt in how they operate and they have become a force to be reckoned with in the autonomous driving industry. In the United States, Google began developing driverless technologies in 2009. From December 2015 to December 2016, the company recorded 635,868 miles of test drives on California roads.[7] It is not only the company with the most mileage in California, but it is also the company with the lowest system deactivation rate—that is, the lowest rate of interference from a human safety driver. Uber, the largest ride-hailing company in the United States, has been approved for unmanned road tests in Pittsburgh, Tampa, and San Francisco, California. In September 2016 the second largest ride-hailing company Lyft announced setting up a three-phase plan for developing self-driving cars, beginning with road tests in Pittsburgh. Apple received its first California test licenses in April 2017. In South Korea, the government has approved technology company Naver to carry out road tests, making it the 13th self-driving company to get a permit. The company plans to make a level 3 autonomous driving car commercial by 2020.

For China, leading technology company Baidu obtained the test license for California in September 2016. Within China, the company first launched unmanned vehicle trials in November 2017 on the roads of Wuzhen, Zhejiang, the site of one of China's largest technology conferences, shortly after Lu Qi, the chief operating officer of the company at the time, announced Baidu's Apollo plan in April of that year. Apollo plans to open up Baidu's autonomous driving technology to other developers and partner companies, making public code and other technologies, including for such functions as environment awareness, route planning, vehicle control, and operating systems. The company also provides a complete set of tools to carry out development testing. The open source approach is intended to further reduce the threshold for autonomous vehicles research and development, so as to facilitate rapid and widespread adoption of the technologies. In the second half of 2016, Chinese Internet

[7] Data comes from Google's annual report provided to California's Department of Motor Vehicles.

conglomerate Tencent established an autonomous driving lab, with a focus on bringing together core technologies for autonomous driving, such as developing 360-degree surround vision, high-precision maps, point cloud (used to represent 3D shapes) data processing, and fusion positioning (combining multiple datasets to aid location).

Startups become targets as acquisitions bring breakthroughs

Companies, often startups, that make rapid progress and take the lead in autonomous driving technologies often quickly become targets for acquisitions. In July 2016, GM acquired Silicon Valley startup Cruise Automation for more than $1 billion, and the RP-1 highway autopilot system developed by the company has the potential for highly automated driving applications. In March 2017, Intel acquired Israeli technology company Mobileye for $15.3 billion, a startup dedicated to developing software and hardware for autonomous driving. The company is the main camera supplier for driver assistance systems in companies such as Tesla and BMW, and holds a number of image recognition patents. In 2015, Uber acquired deCarta, a startup company that builds geospatial software platforms, and also hired a number of Microsoft Bing employees with expertise in image and data collection. In April 2017, Baidu announced a wholly owned acquisition of xPerception, a US technology company that develops visual perception hardware and software solutions, with applications in robotics, augmented reality, and intelligent guides for the blind. The company's machine vision software also allows intelligent hardware to position itself in an unfamiliar environment by calculating an environment's three-dimensional structure and planning paths. Analysts cast Baidu's move as a way to strengthen its visual perception capabilities. Overall, acquiring startups is a war that technology giants have used to help to build competitive advantage and accelerate the accumulation of autonomous driving technology.

When Will Self-driving Cars Be Road-ready?

Despite autonomous driving progressing at breakneck speed, the question remains: when will the technology be truly commercialized and become an integral part of our daily lives? The reality is that there is no way to fully verify the safety of autonomous vehicles before they are widely deployed.

Another key question, therefore, is whether it is necessary for autonomous vehicles to be proven fully safer before being allowed on the road? Even if accident rates for autonomous vehicles may drop far below those for human drivers, many people will likely still resist putting their lives in the hands of a robot that they cannot fully understand.

In May 2017, Nidhi Kalra, an analyst at the RAND think tank, testified to the US House of Representatives on the "Challenges and Processes for Autonomous Vehicle Safety and Mobile Benefits." Kalra highlighted that there are "no demonstrated and accepted methods of proving safety." The crux of the issue, Kalra argues, is that autonomous vehicles need to accumulate rich real-world testing data, making it difficult for a controlled environment to fully simulate the real-world road conditions that are crucial for improving machine learning algorithms. However, the flip side of the argument is that countries are unwilling to allow autonomous vehicles access to public transportation roads without appropriate safety requirements being met, due to the risks for other vehicles, drivers, and pedestrians. In the words of the report, "allowing self-driving cars to get on the road in the real world is like driving a minor to drive a car."

Furthermore, McKinsey's Centre for Future Mobility released a report in May 2017 estimating when autonomous driving robots would be road ready, as well as the commercial deployment time for autonomous vehicles. According to the report, level 4 self-driving vehicle in the SAE grading standard will probably appear in the next five years, while completely autonomous vehicles (level 5) could be over ten years away, due to a large number of current obstacles toward deployment. Level 5 autonomous driving systems are meant to operate the vehicle in any possible environment, but there are many unstructured roads in the real world, without obvious lane marking or road signs, making it much harder for autonomous driving systems. Also, it is hard for software development to keep pace with progress in hardware. The first step is to develop software that can bring together various data needed to identify objects. The relevant data may come from fixed objects in the real word, from point clouds constructed by lasers or from camera images. The next step is then the development of "if-then" instructions that can cover all possible scenarios to simulate human decision-making, a process that requires that data from different scenarios are continuously used to train the artificial intelligence system. The third step is to construct a system of failsafe measures to ensure the safety of passengers in the event of an accident, a process that requires predicting the various consequences that might come from the systems decisions. Building the above software system takes a large investment of time and that is why it will be difficult to create fully autonomous vehicles.

But efforts to deploy autonomous vehicles are not a dead-end. Regulatory authorities can work with companies, research institutions, and universities to find practical and effective ways to test safety, while the methods themselves need to be strictly, objectively, and independently examined and evaluated. Furthermore, there needs to be flexibility for safety test requirements so that autonomous driving companies can have their technology integrated into public transportation systems once it meets requirements. For autonomous driving companies, foundational research on autonomous driving technologies is more effective than hyping of applications. In the future, so long as safety standards are met, technical research can be carried out in the real world with efforts like autonomous driving navigation research in restricted environments and for specific applications; test data, including test mileage, collisions, and system errors, can be openly shared between companies and regulators in order to not only help other companies to reduce research missteps, but also to provide proof of the technology's safety that is necessary for commercialization.

Bibliography

Clark, Joe. "Bosch and NVIDIA take self-driving AI to the next level". 15th March 2017 http://www.cbronline.com/news/internet-of-things/cognitive-computing/bosch-nvidia-take-self-driving-ai-next-level/.

"IBM Patents Cognitive System to Manage Self-Driving Vehicles." https://www-03.ibm.com/press/us/en/pressrelease/51959.wss.

http://www.businessinsider.com/uber-builds-out-mapping-data-for-autonomous-cars-2017-2.

Heineke, Kersten, Philipp Kampshoff, Armen Mkrtchyan and Emily Shao. "Self-driving car technology: When will the robots hit the road?" McKinsey&Company, 2017: 4–10.

Kalra, Nidhi, Challenges and Approaches to Realizing Autonomous Vehicle Safety. Santa Monica, CA: RAND Corporation, 2017. https://www.rand.org/pubs/testimonies/CT463.html.

"Naver gets govt approval for self-driving car road test". 20 February 2017 https://www.telecompaper.com/news/naver-gets-govt-approval-for-self-driving-car-road-test-1184528.

Xiao, Wen. "研发太累了?百度收购科技公司或为抢跑自动驾驶" 14 April 2017. http://www.telworld.com.cn/show-list-7510.html
"福特不甘落后!超10亿美元收购自动驾驶公司". 9 July 2016. https://www.sohu.com/a/102917308_114885
"智能网联汽车标准体系将发布" 19 August, 2016. http://www.iovweek.com/guonei/1883.html.

CHAPTER 6

Intelligent Robots

Robots have long been a staple of science fiction, but now they are quickly becoming a reality. In the mid-twentieth century, the first commercial robot was made in the United States. Today, rapid progress in computers, microelectronics, and other areas of information technology are pushing robotics technology to advance ever faster, while the level of intelligence in robotics is ever higher and the number of applications has greatly increased. Robots can now be found in manufacturing, the service industry, medical care, education, and the military. Combined with people, these robots are changing the world we live in.

What Is a Robot?

A robot is a machine that imitates parts of biological bodies and is able to move automatically, manipulate objects, and perceive its surrounding environment.

There is no uniform global standard for classifying robots, but they can generally be divided by: field of use, specific application, mechanical structure, and method of control. When dividing by field of use, there are two main kinds of robots, namely industrial and service.

In 1987, the International Organization for Standardization defined industrial robots thus: "automatically controlled, reprogrammable multipurpose manipulator programmable in three or more axes."

Further subdivided by specific application, industrial robots can be divided into freight robots, welding robots, assembly robots, vacuum

Tencent Research Institute et al. (eds.), *Artificial Intelligence*,
https://doi.org/10.1007/978-981-15-6548-9_6

robots, palletizing robots, painting robots, cutting robots, and cleaning robots. As service robots are a relatively new classification of robots, there are no particularly strict definitions of service robots, and views on definitions differ between scientists from different countries.

A more recognized definition comes from the International Federation of Robotics (IFR): a robot "that performs useful tasks for humans or equipment excluding industrial automation applications."

China's definition of service robots in the *National Medium- and Long-Term Science and Technology Development Plan (2006–2020)* is: "Intelligent service robots are intelligent equipment that integrate various forms of advanced technology to provide needed services to humans in unstructured environments." Service robots can be subdivided into those for professional and for individual/home use.

The Applications of Industrial Robots Are Maturing and Steadily Growing

Industrial robots are currently the most common kind of data-integrated machines, given their high added value and wide array of applications. As a mainstay technology of advanced manufacturing and an emerging industry for a digitalized society, robotics will in future play an ever-more important role in production and the development of society. Since the global financial crisis, market recovery has brought robotics back to life, with the global industry and market size continuing to grow as governments and multinational companies actively invest in robotics. In 2016, the number of orders for robots globally was 258,900 units, with an inventory of 1,779,000 units. In China 85,000 units were sold compared to an inventory of 332,300 units. The United Nations Economic Commission for Europe (UNECE) and the International Federation of Robotics (IFR) have said that the global robotics market has a promising future, with the industry maintaining a steady growth since the second half of the twentieth century. The robotics industry in Asia grew fastest in that period, reaching a 43 percent increase.

How far has China's industrial robotics industry come since its inception in the 1980s? And what are its main features?

The Market Is Growing Rapidly

Since 2015, China's economy has faced growing downward pressure and its enterprises unexpectedly great challenges. Chinese businesses' demand for automation and intelligent manufacturing equipment, such as intelligent robotics, has grown rapidly in recent years, propelled by the loss of China's demographic dividend and the rapid rise of labor costs.

Chinese Robotics Brands Have Yet to Scale Up

Developing China's domestic robotics industry was made into a clear matter of great importance in May 2015, with the release of *Made in China 2025*, a strategic document released by China's State Council that aims to move its economy up the value chain by producing more advanced products. This roadmap—alongside a plan to bring about "innovative projects to develop intelligent manufacturing equipment" issued jointly by the Ministry of Development and Reform, the Ministry of Finance, and the Ministry of Industry and Information Technology—set out the grand prospects ahead for China's robotics markets, drew international robotics companies into the Chinese market, and intensified domestic competition.

Applications for Robotics Are Ever Growing

Multiple central Chinese government documents—including the Ministry of Industry and Information Technology's guiding document on promoting the "development of the industrial robot industry at the national level," as well as its plans to integrate the raw materials industry and reduce human involvement in explosives production—have brought the use of robotics to various industries. These programs are combined with those of local-level governments to expand the use of robotics from industries like automotics, electronics, metal manufacturing, rubber, and plastics toward others such as textiles, military logistics, explosives, pharmaceuticals, semiconductors, food, and raw materials.

The Areas of China That Are Using Robotics Are Also Ever Growing

In recent years, a great number of Chinese companies, propelled by growing demand and government policies promoting domestic innovation, have begun manufacturing robots either using their own research or by working with research labs. As such, industrial robots in China are now becoming widely used, while service robotics is in an early stage of development. Four industrial clusters for robotics have been formed: around the Bohai Sea in northeast China, the Yangtze River Delta, the Pearl River Delta, and then areas of central and western China.

The Use of Robotics in the Service Industry Is Still in Its Infancy

Service robots appeared later than industrial robots, first appearing in the 1990s. Currently, the use of robotics in the service industry is still in its infancy and has not been fully marketized. But this is starting to change. An aging population and labor shortages, combined with technological advancements, are pushing rapid developments in robotics. The growing number of service robots can be divided up according to their field of application, from personal service robots, such as cleaning, education, and entertainment robots, and professional service robots, such as those used in defense robots, medical robots, and logistics robots. At this time, more than 20 countries in the world are developing service robots. In this field, the United States, Germany, and France are the western leaders, with Japan and South Korea being at the forefront of development for Asia.

With the arrival of an artificial intelligence era, developed nations have made service robotics part of national growth strategies. China is no different and has released numerous related policies, making the use of service robots a strategic technology that must be developed as a priority in the future. With a particular focus on high-end intelligent equipment, China plans to develop and cultivate a core of service robotics enterprises with a total output of RMB 10 billion. The highest priority has been given to developing robots for use in public security and medical treatment, as well as to building capacities in bionic robotics and modular core robotics components. With strong policy support, China's service robot industry is expanding rapidly. As it stands, China's service robot market has the following features:

Low Market Penetration Rates

Due to China's service robot industry's late beginnings, combined with low spending by Chinese consumers, the penetration rate of service robots in China is relatively low. Service robots began to be sold at scale in 2005. Chinese companies that design and produce service robots mostly sell in the low-end market, a significant difference from developed markets like Japan and the United States. Currently, products that have begun to be industrially produced at scale include cleaning robots, as well as educational and entertainment robots.

Applications Are Becoming Ever More Mature

Service robots now have applications in individual or family services, medicine, and the military, as well as in a number of other specialized fields. Smart home, entertainment, education, safety, health, and information services all are individual or family uses. This is the most mature and competitive area of the service robotics. The underlying technology for individual or family uses is relatively simple, the use of it is clear, and it is easily commercializable. Several companies in China are currently developing such robots. Within medical robots there are surgical robots and rehabilitation robots. Medical robots in China are still in an early, gradual stage of development. Due to a lack of relevant expertise and technology, China's gap with developed countries in the field of medical robots is huge. There are currently no mass-produced medical robots in China, and their use in major hospitals remains infrequent. Military robots can be divided into those for are aerial, underwater, and space. China's military robots currently remain in an early stage of development. In June 2014, SIASUN Robot & Automation Co., a robotics company under the Chinese Academy of Sciences, became the first robotics company to be approved as a level 1 supplier of military equipment, making it qualified to supply certain military equipment classified as confidential at the second degree, as well as computer system integration classified as confidential at the first degree.

Furthermore, China also has a large demand for robots with applications in public security, agriculture, surveying, and mapping. Some listed companies in China have already begun to sell robots that can help with police patrols or firefighting with some early successes. For agriculture and mapping, the level of machine use in China's agricultural industry remains low, so on-the-ground agricultural robotics technology is still in early stage of research, while mapping also lacks sufficient support from advanced technologies. Drones, however, are more easily commercializable, have a number of potential applications, and have good prospects for a growing market.

Positive Prospects for Market Growth

China faces far fewer obstacles to develop service robots than industrial robots. The gap between Chinese companies and their foreign counterparts is relatively small in the field. This is because service robots are often developed for specific markets, which allows Chinese companies that work closely with other local companies an advantage over international competitors in the China market. Additionally, service robots are still an emerging industry internationally, with most large companies in the field only five to ten years old, meaning that a large number of players remain in early stages of development, giving Chinese companies an opportunity to narrow the gap. Furthermore, service robots are focused on consumers, meaning the market is vast. Factors like an aging population and a sharp rise in labor costs are likely to spark a blossoming of the service robots market.

Trends in International Robotics Development

All the world's largest industrial powers have policies for the robotics industry, such as Germany's Industry 4.0 program, Japan's *New Robot Strategy*, and the US Advanced Manufacturing Partnership. These plans are an important part of the development of the robotics industry and will not only promote the continued growth of industrial robotics but will also spur rapid growth for the professional and individual use of service robots.

The automotive industry is a major user of industrial robots. Automakers are currently still the largest users of industrial robots, with the ratio of industrial robots to human employees standing at 1 to 1000 in Japan and Germany.

The dual-arm cooperative robot is a new bright spot in the industrial robot market. With the continuous increase of labor costs, the labor cost burden of large assembly factories and SMEs relatively heavy, the serious ageing of the population and the shortage of labor in the country, the dual-arm cooperative robot is a solution to reduce labor costs, improve production efficiency and make up for the labor gap.

The growth momentum of the service robot market is very promising. At this stage, this market is mainly about sweeping robots, entertainment robots and medical care robots. In addition, the aging of the agricultural population in some countries and regions is becoming more and more serious, which will also drive the demand for agricultural robots.

Trends in the Development of China's Robotics Industry

China's research and development has been becoming stronger. China's industrial robots started late, and, while some powerful model companies have appeared as the industry begins to take shape, the Chinese industry's overall ability to innovate in robotics for the most part clearly lags behind advanced manufacturers in other countries. For China to make technical breakthroughs that will reduce costs, as well as to move up the value chain to mid- to high-end manufacturing, Chinese companies will need to strengthen their expertise and invest more time and energy in research.

Nothing can prevent an intelligence upgrade for manufacturing. Growth in China's labor force is slowing at the same time as the relative non-working population is rapidly growing. A shortage of labor is coming as the country's demographic dividend disappears. The most effective way to resolve this problem is to automate and upgrade China's manufacturing industry. Strong support from the Chinese government and the pressure of transformation of traditional industries will continue to drive a lively interest in robotics and growing market participation in the industry.

Service robots will either catch up or overtake humans. Aging populations, labor shortages, and growing demand from across society makes the widespread adoption of service robots inevitable. In this emerging industry, the gap between China's development and developed nations is small and Chinese companies can find competitive advantages through their understanding of the local market and culture. As such, there is much greater opportunity in the service robotics industry for a manufacturer which takes a leading position to corner a huge chunk of the market.

Policy support will become more standardized and refined. China's robotics industry has attracted a lot of capital due to the favorable government policies and belief in the market's huge potential. As such, there are now dangers of the market becoming overheated. To avoid the blind expansion of the robotics industry and prevent high-end production from becoming low-quality, the Chinese government will need to continue to standardize and refine its support, so as to ensure an orderly and healthy development of China's robotics industry.

Bibliography

Bar-Cohen, Y. and Hanson, D. *The Coming Robot Revolution: Expectations and Fears About Emerging*. New York: Springer, 2009:8–9.

"2016机器人产业发展分析与展望". 24 March 2016. 1 June 2017. http://www.cnelc.com/Article/1/160324/AD100344725_1.html.

"2016年全球机器人和"工业4.0"市场趋势分析". 7 April 2016. 22 April 2017. http://www.gongkong.com/news/201604/340784.html.

"2016年中国机器人行业发展趋势预测:智能化+服务化". 20 July 2016. 22 April 2017. http://www.globalrobot.com.cn/news/2/_7633.html.

"2016中国服务机器人产业发展白皮书". 4 January 2017. 1 June 2017. http://robot.ofweek.com/2017-01/ART-8321203-8100-30087531.html.

"我国工业机器人产业发展战略与对策研究". 11 May 2015. 22 April 2017. http://www.360doc.com/content/15/0511/15/2584926_469690990.shtml.

"国务院关于印发《国家中长期科学和技术发展规划纲要 (2006—2020年)》的通知". 27 May 2017. http://www.gov.cn/zwgk/2006-02/14/content_191891.htm.

Xu, Fang. "发展我国工业机器人产业的思考".*机器人技术与应用*. 5(2010):5–6.

CHAPTER 7

Smart Healthcare

In recent years, the development of smart healthcare has been heating up. Some people suggest that although security smart investment receives the most attention, artificial intelligence might first be used in the field of medicine. According to a CB Insights report on the state of artificial intelligence from August 2017, medicine and healthcare was the field that had invested most in artificial intelligence with over 270 deals having already been done since 2012. Breakthroughs in key technologies like image recognition, deep learning, and neural networks have brought about a new phase of artificial intelligence, which has driven deeper integration of artificial intelligence into a medical industry that is now more data-intensive and intelligence-driven. At the same time, society is increasingly aware of health issues and the population is aging, creating a pressing demand for improved medical technology, extended human life, and greater overall health. The industry also faces problems like the uneven distribution of medical resources, a lengthy development cycle for pricey drugs, and the high costs of training medical personnel. Such a clear demand for medical progress has sparked a wave of innovation that has sought to upgrade the medical industry using artificial intelligence.

Core Smart Medical Uses

Looking at the global activity of startups, smart healthcare is being used for oversight and risk management; medical research; medical imaging, and diagnosis; lifestyle; mental health; nursing; the management of

Tencent Research Institute et al. (eds.), *Artificial Intelligence*,
https://doi.org/10.1007/978-981-15-6548-9_7

hospitals or emergency rooms; drug discovery; virtual assistants; and wearables, among others. Overall, artificial intelligence in healthcare is mainly used in the following five areas:

Medical Robots

It is not uncommon for robots to be used in medicine. For example, there are a number of technologies, including smart prosthetics, exoskeletons, and other auxiliary equipment, that help to repair severe injuries, as well as healthcare robots that assist medical staff. There are currently two main kinds of medical robots in use. One is a wearable suit known as a "smart exoskeleton" capable of reading signals from a human body and brain to aid movement and help avoid injury. The second is a robot that can conduct surgery or other healthcare functions, a typical example of which is Da Vinci Surgical System, developed by American company Intuitive Surgical.

Smart Drug Development

Smart drug development refers to the use of deep learning to drug research, using big data to rapidly and accurately unearth and select suitable chemical compounds or organisms, so as to shorten the development cycle of new drugs, as well as reducing costs and improving the success rate. Artificial intelligence can use computer simulations to predict a drug's activity, safety, and side effects. Using deep learning, new breakthroughs have been made in cardiovascular and antineoplastic drugs, as well as those that treat common infectious diseases. The technology has also played an important role in developing drugs to fight the Ebola virus.

Smart Diagnosis and Treatment

Smart diagnosis and treatment is the use of artificial intelligence to assist in diagnosis and treatment. By "teaching" the computer the medical knowledge of an expert doctor, such systems can simulate the doctor's diagnostic reasoning and thus reach a diagnosis and produce a reliable treatment plan. Smart diagnosis and treatment is the most important use of artificial intelligence in healthcare. Artificial intelligence can process vast amounts of data faster and use deep learning to discover patterns before summarizing regular discrepancies so as to produce a diagnosis for the sickness.

Intelligent Medical Imaging

Intelligent medical imaging involves using artificial intelligence in the diagnosis of medical images. There are, broadly speaking, two parts to this process: one is image recognition during the perception stage, where the main purpose is to analyze the image and find significant information. The second is to use deep learning to study and analyze images by inputting a large amount of image and diagnostic data in order to allow the neural network to develop diagnostic ability through continuous training.

Intelligent Health Management

Intelligent health management applies artificial intelligence to specific aspects of health management, with a current focus on using accurate medical science for management, as well as in risk identification, virtual nurses, mental health, online consultation, and health interventions.

(1) Risk identification: extracting and analyzing health data using artificial intelligence technology to identify the risk of a disease occurring and then offer measures to reduce risk.
(2) Virtual nurse: digital avatars that collect data on personal habits like eating habits, exercise, and medication then use artificial intelligence to assess the overall healthiness of the patient's lifestyle so as to help plan daily activities.
(3) Mental health: using artificial intelligence to recognize emotions using an individual's language, expression, and voice data.
(4) Mobile medical care: combining care with artificial intelligence to provide mobile medical services.
(5) Health intervention: using artificial intelligence to gather physical data from a user and then create a customized care plan.

Examples of Smart Medicine Applications

Medical Robots

One kind of medical robot is the smart exoskeleton. Russia's ExoAtlet makes two smart exoskeleton products—ExoAtletI and ExoAtlet Pro—the first for individual use and the latter to use in hospitals. ExoAtletI is suitable for patients whose bodies are paralyzed below the waist. So long as the individual retains near-complete upper-body function, the exoskeleton can help patients walk, climb stairs, and carry out some specialized training movements. ExoAtlet Pro adds additional features on top of those of ExoAtletI, such as the ability to measure pulse, to stimulate muscles using electricity, and to set up a fixed walking pattern, among others. Japan's Ministry of Health, Labor and Welfare has officially listed "robot suits" and "medical hybrid auxiliary limbs" as medical devices when sold in the country, with the devices being used to improve the walking ability of patients with amyotrophic lateral sclerosis (or ALS), muscular dystrophy, and other similar conditions.

A second kind is surgical robots. A representative example of this kind of robot globally is the Da Vinci Surgical System. The Da Vinci Surgical System can be divided into two parts, the first being a terminal beside the operating table that allows remote control of robot arms by a doctor. The system allows the surgeon to operate on the patient using robot arms that are far more flexible than a human's. The system also includes a camera that can enter the body during surgery, allowing operations with only very small incisions and those that would be difficult for a human to complete. From this control terminal, the computer uses several cameras that restore the two-dimensional image from different angles into high-definition three-dimensional images of the body, allowing the doctor to monitor the entire surgical procedure. Thousands of Da Vinci robots have been assembled worldwide and they have been used in millions of operations.

Smart Drug Development

Efficiency is the key to drug development. San Francisco–based Atomwise, for example, has partnered with IBM to analyze compounds that could be potential cures for diseases by comparing vast numbers of molecular structures and their interactions against those that have been successful cures in the past. In 2015, Atomwise was able to recommend two potential drug candidates to reduce Ebola infectivity in less than a day of computation by simulating the impact of 7000 existing drugs on the viruses' mechanism for entering cells. In addition to unearthing compounds to develop new drugs, US pharma company Berg is developing new drugs by studying biological data. Berg uses an artificial intelligence platform called

"Interrogative Biology" to carry out research into basic structure of human health, molecules of the body, and cell defense systems, as well as mechanisms of pathogenesis, by using artificial intelligence and big data to work out and tap into the naturally existing disease-fighting capacity of molecules within the human body. This approach could cut in half the time needed to develop new drugs to fight intractable diseases like diabetes and cancer.

Intelligent Diagnosis

The earliest international application of artificial intelligence to medical diagnosis was the MYCIN expert system in the 1970s. In China, an artificial intelligence–powered expert system was also developed in the 1970s, but it has developed rapidly since. One of China's first systems was the "Guan Youbo Hepatitis Medical Expert System" developed by Beijing College of Traditional Chinese Medicine. The program simulated the diagnosis procedure for liver diseases used by the famous Chinese doctor Guan Youbo. In the early 1980s, Fujian College of Traditional Chinese Medicine and Fujian Computer Center developed another diagnosis system for bone injuries named after Lin Rugao, another well-known doctor. Institutions including Xiamen University, Chongqing University, Henan Medical University, and Changchun University, among others, have also developed artificial intelligence–based medical computer expert systems, all of which have been successfully used in clinical practice.

IBM Watson is currently the most mature example of intelligent diagnosis. In 17 seconds, IBM Watson can read 3469 medical monographs, 248,000 academic papers, 69 treatment plans, 61,540 trial data, and 106,000 clinical reports. In 2012, Watson passed the United States Medical Licensing Examination and was deployed as a supplementary medical device in a number of hospitals in the United States. At present, Watson can help with the diagnosis of a variety of cancers including breast cancer, lung cancer, colon cancer, prostate cancer, bladder cancer, ovarian cancer, and uterine cancer. Watson is an artificial intelligence system that combines natural language processing, cognition technology, automated reasoning, machine learning, and information retrieval, among other technologies, in order to produce supposed understanding by collecting a lot of evidence to analyze and evaluate.

Intelligent Image Recognition

The artificial intelligence system developed by the Beth Israel Deaconess Medical Center (BIDMC) at Harvard Medical School can accurately detect cancer cells in images of lymph nodes 92 percent of the time. American startup Enlitic uses deep learning to detect malignant tumors like cancer. Enlitic's system detects cancer at a rate above that of a panel of top radiologists and was shown to find 7 percent more cancers human doctors failed to spot.

Intelligent Healthcare Management Systems

The first such system is used to identify risks. Lumiata provides predictive risk analysis using its core product Risk Matrix to map the risk of disease over time to users based on data from a large number of health plan members or patients' electronic medical records and pathophysiology. It uses the medical graph analysis to make a rapid, targeted diagnosis of the patient, thus reducing the patient's triage time by 30–40 percent.

The second kind of system is virtual nurses. US-based Verint Next IT has developed an application called Alme Health Coach that is configured for specific diseases, drugs, and treatment plans. It can be synchronized with the user's alarm clock in order to answer particular questions such as "how am I sleeping." The app can also prompt users to take medicine on time. This idea is to collect actionable data that can then be made available to a doctor to help them better interact with a patient. The app is mainly for patients with chronic diseases. By integrating data from sources such as wearable devices, smartphones, and digital medical records, the app comprehensively evaluates the patient's condition and provides a personalized healthcare management solution. The US National Institutes of Health (NIH) has invested in an app by a New York–based startup AiCure that automatically works out whether a patient is taking the correct medications using AI to analyze photos of patients taken with a smartphone camera.

The third kind of system is for mental health. In 2011, US-based Ginger.io developed an analytics platform to track any weakening of its users' mental health using smartphone data to see whether the users' habits have changed, as well as actively asking users questions about their usual behavior. If there is a significant change in user behavior, the user will be notified, as will their in-app therapists and coaches. Another kind of emotional recognition technology developed by a US-based company called Affectiva uses webcams to capture and record users' expressions so as to work out whether they are feeling emotions like joy, disgust, or confusion, with potential applications from branding to political messaging.

The fourth kind of system is mobile medical care. London-based Babylon's online medical treatment system uses artificial intelligence combined with users' responses to provide initial diagnosis and specific treatment recommendations based upon symptoms listed in the user's medical history. AiCure is a smart healthcare company that reminds users to take medication on time, using mobile technology and facial recognition to determine whether patients take medication on time, then use patient data in the app with automated algorithms to identify medications and drug intake.

The fifth kind of system is health interventions. Denver-based Welltok uses artificial intelligence to analyze user data, including that provided by wearables partners like Map My Fitness or Fitbit, to provide personalized lifestyle interventions and preventive health management programs.

Chinese Smart Healthcare Development

According to an in-depth report on Internet medical treatment released by China's Founder Securities, "The development of China's Internet medical treatment has gone through three stages: a stage of information services, connecting people with information; a consultation service stage, connecting people with doctors; and a diagnosis stage, connecting people and medical institutions." China's smart healthcare industry is still in its infancy, but capital has been flooding to smart healthcare startups. Looking forward, Chinese companies should strengthen research and investment at the foundational level of databases, algorithms, and other general-use technologies, so as to build a strong foundation while at the same time expanding the practical applications of smart healthcare.

Challenges Facing Smart Healthcare

Government regulatory barriers: the medical field is one of the most highly regulated sectors, so adding artificial intelligence and smart healthcare products to the industry will bring about adjustments of government regulations. An in-depth advancement of smart healthcare will need to first meet regulatory requirements. China's General Office of the National Health Commission, a top regulatory body, in late 2018, released trial documents on the *Administrative Measures on Internet-plus Healthcare* and *Guiding Opinions on Promoting the Development of Internet Medical Services*, which will strictly supervise Internet diagnosis and treatment service, emphasizing that quality and safety must be guaranteed in accordance with Chinese law.

Another challenge is the limited access to medical data. Due to the sensitive nature of medical information like data about a person's genes or their illnesses, many countries have put in place stricter regulations regarding the collection, storage, and use of medical data than for that of more general-use data. In Australia, for example, medical data cannot be sent outside the country, unless there are extraordinary circumstances. The United States also requires that the commercial use of medical information must strictly comply with the Health Insurance Portability and Accountability Act and the Health Information Technology for Economic and Clinical Health Act. Smart healthcare is dependent upon the accumulation of a large amount of medical data for rapid growth. As Rajeev Ronanki, head of Deloitte Healthcare, said, only a combination of three forces can drive machine learning forward: exponential data growth, faster-distributed systems, and algorithms that more intelligently handle and understand data. Limitations on the access and use of medical data will also to a degree hinder smart healthcare's development.

A final challenge is the difficulty of becoming compatible with traditional hospitals. Although many hospitals now have a level of informatization, promoting smart healthcare requires hospitals to fundamentally update their IT systems and information services. This involves not only updating equipment and facilities, but also training relevant personnel, which can be costly and clash with more traditional business models. Potential incompatibility with traditional services will also slow the development and adoption of smart healthcare.

CHAPTER 8

AI-powered Investment Advice

AI-powered investment advice comes from integrating artificial intelligence deep within financial services. This method of investment began in 2008, and since 2011, the market for using AI in finance has accelerated significantly in the United States and elsewhere. Using AI in investment management has been a major breakthrough for wealth management; as compared to traditional models, AI-powered investment advice facilitates higher transparency, lower investment thresholds, and reduced management costs, as well as a better user experience and personalized investment advice, all of which are unique advantages attractive to certain kinds of customers, who have driven overall market growth as more users adopt such techniques.

What Is AI-powered Investment?

At this time, there is no authoritative, unified definition—either theoretically or in practice—of what constitutes AI-powered investment advice (also sometimes known as robo-advisor), which has been adopted to varying degrees in businesses in different countries. Other ways of referring to AI-powered investment advice include: "automated advice tools," "automated investment platforms," and "automated investment tools," among other terms. Wikipedia defines robo-advisors as a type of financial advisor that provides financial advice or online investment portfolio management services using algorithms to allocate, manage, and optimize a customer's assets, thereby minimizing human intervention.

Tencent Research Institute et al. (eds.), *Artificial Intelligence*,
https://doi.org/10.1007/978-981-15-6548-9_8

In May 2015, the US Securities and Exchange Commission (SEC) and the Financial Industry Regulatory Authority (FINRA) jointly issued investor tips for the use of automated investment tools, which broadly defined AI-powered investment tools for businesses: "These tools range from personal financial planning tools (such as online calculators) to portfolio selection or asset optimization services (such as services that provide recommendations on how to allocate your ... brokerage account) to online investment management programs (such as robo-advisors that select and manage investment portfolios)." In February 2017, the SEC issued a smart investment guideline, which said: "Robo-advisers, which are typically registered investment advisers, use innovative technologies to provide discretionary asset management services to their clients through online algorithmic based programs. A client that wishes to utilize a robo-adviser enters personal information and other data into an interactive, digital platform (e.g., a website and/or mobile application). Based on such information, the robo-adviser generates a portfolio for the client and subsequently manages the client's account."

In August 2016, the Australian Securities and Investments Commission (ASIC) issued its Regulatory Guide 255 *Providing Digital Financial Product Advice to Retail Clients*. The guide uses the term "digital advice" to refer to "the provision of automated financial product advice using algorithms and technology and without the direct involvement of a human adviser. It can comprise general or personal advice, and range from advice that is narrow in scope (e.g. advice about portfolio construction) to a comprehensive financial plan."

In December 2015, Europe's three major financial regulators—the European Banking Authority (EBA), the European Securities and Markets Authority (ESMA), and the European Insurance and Occupational Pensions Authority (EIOPA)—jointly issued a joint discussion paper on automation in financial advice, which analyzed emerging trends in digital and automated insurance, banking, and securities: Automation in relation to financial advice is a more mature phenomenon ... In this business model, automated tools are used as a type of financial adviser, often referred to as a "robo-adviser": the automated tool asks prospective investors for information about their specific circumstances and, based on the answers provided, an algorithm is used to recommend transactions in financial instruments that match the customer's profile. Different automated tools may be used to support different parts of the advice process, for example the collection of information, risk profiling, portfolio analysis, and order processing or trading. As such, AI-powered investment can be understood as a combination of multiple automation tools.

In September 2015, the Canadian Securities Administrators (CSA) issued guidelines placing specific requirements on portfolio managers which provide online investment advisory services. It is worth noting that the CSA guidelines focused on the regulation of registered portfolio managers (PMs) and advising representatives (ARs) that were providing investment management services to retail investors through an interactive website. As such, the online investment consultants were not "AI-powered investment." They are rather using a web platform to improve the efficiency of their services and provide customers with investment advice. As such, human consultants still actively participate and take responsibility for the customer's decision-making process when investing.

In China, a formulation used by authorities that is close to AI-powered investment advice comes from *Provisional Regulations on Strengthening Supervision of the Use of "Stock-picking Software" in Securities Investment and Consulting Businesses* issued in 2012 by China's Securities Regulatory Commission (CSRC), which defined such software as software or equipment that provides investment consulting services. Such services include providing investment analysis concerning specific classes of stocks; forecasting prices of particular securities investment products; suggesting particular securities investment products; and providing advice on when to buy and sell specific securities investment products; as well as providing other kinds of analysis, forecasts, or recommendations for securities investment.

According to article two of the provisions, selling or providing "stock-picking software" to investors in order to directly or indirectly receive economic returns is a form of securities investment consulting business and should obtain a relevant business license from the CSRC. At a press conference in August 2016, CSRC further clarified that AI-powered investment advice is still in essence an investment advisory service with the same basic format, meaning that practitioners and institutions must have the relevant qualifications and business licenses.

Factors Leading to the Rise of AI-powered Investment Advice

There are a number of diverse factors behind the rise of AI-powered investment advice. Overall, using innovative technologies can satisfy market demand that traditional investment models fail to meet, while changes in investor habits and social backgrounds, as well as changes to market structure and the regulatory environment, have pushed ever-greater penetration and development of AI-powered investment advice.

Relatively high capital thresholds and fees for traditional investment services mean that the financial management needs of low-net-worth clients are not effectively met. In the United States, for example, the average capital threshold for traditional investment services is $50,000 and the average management fee is 1.35 percent of the assets under management. For AI-powered investment advice, the minimum capital threshold can be as low as $500, with fees between 0.02 and 1 percent. Some AI-powered investment advisors do not set a minimum threshold for capital investment and do not charge management fees.

Target users for these services are generally in the long tail of investors, those with less than $200,000 in capital. The management fee is between 0.3 and 0.5 percent, which is 60 to 70 percent of the fees of human advisors.

Changes in investor habits have also sparked a new demand for different kinds of investment and wealth management services. Younger users born in the 1980s and 1990s prefer communicating online and want to access personalized services anytime, anywhere. The non-face-to-face, customized model provided by AI-powered investment advisors is more suited to these users' investment and financial management needs. Millennials born between 1980 and 2000 have already become the core user group for AI-powered investment advisors in the United States, Australia, Canada, and other markets.

From the supply side, technologies like big data, cloud computing, and artificial intelligence reaching relative maturity have laid a foundation for AI-powered investment advisors. There is much more diverse and rich data today, due to the penetration of the Internet and mobile combined with company efforts to mine and analyze big data, while data processing costs are ever lower. Together these factors allow AI-powered investment advisors the potential to offer more accurate and personalized investment advice and portfolio plans.

Technological development has also lowered the threshold for startups and technology companies become wealth managers. Many such entities have entered the wealth management field by leveraging their technological advantages. At the same time, commercial banks, insurance companies, and large wealth management institutions have integrated AI-powered investment advisors into their consulting business, so as to extend and expand the means and capacities of their traditional service business. By taking advantage of the large customer base and established brands, these institutions have supported the popularization of AI-powered investment advisors.

The Business Model of AI-powered Investment Advice

Different kinds of AI-powered investment advice deploy different investment concepts, methods, and strategies. There is also great variance in the complexity of algorithms that underpin these systems, from simple algorithms that construct a single portfolio to algorithms with multiple strategies that assess thousands of different financial tools and scenarios. Here, we will briefly analyze the business models for AI-powered investment advice by looking at the systems' functions, the target they provide services to, the institutions that use them, and the degree of human participation in the process.

Functions

In March 2016, the US Financial Industry Regulatory Authority released the *Report on Digital Investment Advice*, which gives a detailed description of the functions of digital investment advice. Such tools "support one or more of the following core activities in managing an investor's portfolio: customer profiling, asset allocation, portfolio selection, trade execution, portfolio rebalancing, tax-loss harvesting and portfolio analysis. These investment advice tools can be broken down into two groups: tools that financial professionals use, referred to here as 'financial professional-facing' tools, and tools that clients use, referred to here as 'client-facing' tools."

There are also companies that execute transactions on the basis of AI-powered investment in Canada, Australia, Japan, and Hong Kong, among others, with deals carried out either through their own brokers or via external partners. The United States uses a "dual registration system," whereby the combination of AI-powered investment advice and transaction execution requires qualifications as both an investment consultant and as a brokerage. Some providers of AI-powered investment advice can also provide tax optimization plans, which are widely used in the US market. These providers allow tax optimization by rationally allocating taxable assets, adjusting asset allocations, and offsetting part of asset returns to reduce capital gains tax and allow tax incentives.

The Target for Advice Services

AI-powered investment advice can be divided into business-to-consumer, for ordinary individual investors, and business-to-business, for institutional investors. In the B2B model, the platform provides specialized services, including technology, asset allocation, and risk management, for institutional investors or traditional investment entities, which help to cut costs from traditional investment models and grow business.

Service Providers

Different types of market participants are actively involved in the field of intelligent investment advisory services. Startups that developed proprietary algorithms for automated investment advice models face the challenge of attracting clients from scratch, but these institutions have the advantage of being only lightly regulated in some countries that have yet to incorporate innovative technology companies into their regulatory frameworks. Traditional wealth management entities such as commercial banks and asset managers have also begun to provide AI-powered investment advice to customers. These institutions can leverage their brand and large customer base to reduce costs for their services, improve user experience, and expand the investment services they offer, which in turn can tap into the retail investor market.

Degree of Human Participation

Some AI-powered investment advice platforms are fully automated, and there is little or no human intervention in the process. Others adopt a hybrid model, combining human and machine. During the crucial stage of business development, humans are required, but the approach and degree of this involvement can vary. For example, in the US AI-powered investment advice model, humans provide technical and customer support, as well as talking to customers about specific investment recommendations. Due to regulatory requirements, Canada allows only mixed human-machine services. Some countries, including Australia, Germany, and the United Kingdom, require AI-powered investment advice platforms to set up corresponding safeguard mechanisms that ensure the appropriateness of investment recommendations. Australia and the United Kingdom have also proposed restrictions on investment advice provided by fully automated systems. Most fully automated platforms give customers the option of contacting professional consultants via online chat, phone, or video calls.

Customization

Some AI-powered investment advice companies set asset allocation portfolios in advance, using information provided by users to classify investors and select the appropriate portfolio. There are also some platforms that offer a more personalized or customized asset allocation plan that can optimize existing portfolios in line with the investment objectives and risk appetite specified by the user.

Investment Products

In the majority of countries, the most common investment products provided by AI-powered investment advice companies are investment funds, mutual funds, and exchange-traded funds (ETFs). Some companies in Brazil provide fixed-income products such as government bonds. In countries like Germany and Australia, available products include bonds and stocks. The portfolio schemes recommended by providers in the United States, France, and Turkey include some over-the-counter products such as financial derivatives (contract for difference, or CFDs), foreign exchange, and binary options. In these countries, even bitcoin has become a potential investment product.

Market Size

According to incomplete statistics, there are nearly 140 institutions offering AI-powered investment advice globally, with more than 80 being established after 2014, in markets including the United States, Europe, Australia, India, Canada, and South Korea, among others. The varying definitions of AI-powered investment advice in different countries mean that there are no official statistics on the overall size of the global market. By compiling data from a number of consultancies, we can see that AI-powered investment advice on accounts for a small portion of assets under management compared to traditional investment institutions. In 2015, the net global assets of public funds managed by traditional asset management institutions was $37.19 trillion, while the AI-powered asset management was about $600 billion. The market for AI-powered advice is growing rapidly, however. According to data from BI Intelligence, by 2020 10 percent of the world's assets will be managed by AI-powered investment advisors, with the market size reaching about $8.1 trillion, within which the Asian market will reach $2.4 trillion and the US market will reach $2.2 trillion.

The United States was the first market to have an AI-powered investment advice model. In addition to independent companies like Wealthfront and Betterment, some traditional financial institutions such as Vanguard and Charles Schwab have also launched their own AI-powered investment advice businesses, while manager BlackRock acquired the startup Future Advisor to officially enter the market. Merrill Lynch, Wells Fargo, and other commercial banks have also begun to make forays into the market.

In September 2016, traditional US asset managers had about $52 billion managed by AI-powered investment models, with an average annual compound growth rate of 179 percent. In the same period, assets under management by US independent enterprises using AI-powered platforms grew by 56 percent, reaching $13.2 billion. In Australia, startups and traditional asset management institutions are actively using AI-powered investment models, as a new frontier for the Aus$2.3 trillion pension market. In Europe, AI-powered investment is still at an early stage. AI-powered investment advice is being used in asset management in the United Kingdom, Germany, and Italy. In the Chinese market, according to incomplete statistics, there are more than 20 financial platforms that offer "AI-powered investment advice," including established financial institutions like Ping An Insurance, China Merchants Bank, Minsheng Securities, and GF Securities that have integrated AI-powered investment advice into their systems in order to offer new functions and increase attractiveness to investors. At the same time, Internet finance companies such as Jingdong Finance, as well as independent platforms such as Sparanoid, Hithink Flush Information Network Co, and Clipper Advisor, are cooperating with brokerages both in China and abroad to expand the market for AI-powered investment advice.

CHAPTER 9

Smart Homes

In recent years, the smart home industry has been rapidly developing. With the empowerment of artificial intelligence technology, the industry ecology has gradually improved and matured and the bright future of new smart living is looking to become a possibility. In general, the global development trend of the smart home is good and the United States is leading this industry development trend.

Smart Homes Are Displaying Strong Vitality on the Global Scale

Market research consulting firm Markets and Markets recently announced a report indicating that the global smart home market will reach $122 billion by 2022 and the average annual growth rate between 2016 and 2022 is expected to be 14 percent. According to market research company GfK's research report on seven countries, over half the respondents believed that in the next few years smart homes will have an impact on their lives. Published national data indicates that more than 50 percent (51 percent) of users expressed an interest in smart homes, which is on par with interest in mobile payments and far exceeds interest in wearables and other options (33 percent). With the development of 5G and other next-generation mobile communication technologies, artificial intelligence technologies such as speech recognition and deep learning have also

Tencent Research Institute et al. (eds.), *Artificial Intelligence*,
https://doi.org/10.1007/978-981-15-6548-9_9

matured. Under the integration of new technologies with smart homes, product categories will increase, system ecologies will mature, and the user market penetration rate will increase as a trend of the times.

The United States Is the Sole Leader Spearheading the Industry Development Trend

According to statista's figures, in 2016 the US smart home market had a capacity of $9.7125 billion, making it the country with the largest capacity in the global smart home market. The second to fifth positions were held by Japan, Germany, China, and the United Kingdom respectively. In addition, looking at the growth of the smart home penetration rate the United States also ranked first with 5.8 percent, while Japan, Sweden, Germany, Norway, and other traditionally developed countries ranked second to fifth, respectively, and China's penetration rate was only 0.1 percent. In recent years, the US smart home market capacity has increased by an average annual growth rate of $3 billion, displaying a rapid upward trend. Moving from the general to the specifics, taking 2014 as an example, the capacity of the US smart home industry mainly covered five major domains including entertainment ($13.321 billion), security ($8.368 billion), automation ($76.99 million), energy management ($382.6 million), and environmental cleanliness ($0.829 billion). In 2015, the growth rate was led by automation (147.3 percent) and environmental cleanliness (92.2 percent).

China's Potential Room for Expansion Is Enormous, Is the Market Window About to Arrive?

According to a report released by the Prospective Industry Research Institute, the size of China's smart home market reached RMB60.57 billion in 2016 (statistical results among different organizations also differ). The industry report published by Qianjia Consulting Co., Ltd. divided the development of smart homes in China into four stages, namely the germination period (1994–1999), the creation period (2000–2005), the volatile period (2006–2010), and the integration and evolution period (2011–2020). Qianjia Consulting believes that China is currently at the fourth stage. Since 2011, the smart home market entered a stage of rapid development and the integration of AI technologies has spawned a large number of new technologies, new models, and new businesses and created a huge market demand. Giant companies in China are rushing to launch new products and carve up this huge consumer market. The industry's landscape has evolved, with protocols and technology standards trending toward increasing interconnectedness and new products continuously emerging. With this round of exploration, the arrival of the market's outbreak may just be a matter of time. Aowei Consulting (AVC) expects that by 2020, the penetration rate of smart television will reach 93 percent and the penetration rates of smart washing machines, smart refrigerators, and smart air conditioners will increase to 45 percent, 38 percent, and 55 percent respectively. It can thus be seen that the trend toward the intelligentization of homes is irreversible. The potential market size of China's smart home industry inspires the imagination and is expected to become the market's next window of opportunity.

In Competition Over Smart Homes, Leading Companies Are Raring to Go

The development prospects of smart homes have attracted the involvement of many giant companies, creating a landscape of competition. Domestic and foreign technology companies are eager to try the smart home market, hoping to mass-popularize a single product as a foothold to seize dominance of the smart home industry. As major business players enter the game, the smart home industry ushered in an activation period, paving the way for rapid development.

From an international perspective, companies like Amazon, Apple, and Google are vying to set up strategic footholds on platforms and systems with the intention of using the open platform as a selling point to construct an open ecology. This will realize the strategic goal of an interconnected smart home center while seizing more upstream and downstream holder resources to secure their own market dominance.

Facebook Releases Artificial Intelligence Butler Jarvis

In December 2016, Zuckerberg revealed the company's newly developed artificial intelligence butler "Jarvis." This butler can not only regulate indoor environments, schedule meetings, make regular breakfasts, automatically wash clothes, identify, and entertain visitors, but even teach Zuckerberg's daughter Mandarin.

Google Releases Google-Home and Regroups Nest

On May 19, 2016, Google launched a brand-new Google-Home smart speaker at the 2016 Google I/O Developer Conference. At the end of August of the same year, the entire platform team of smart home company Nest Labs was reconstituted as part of Google in order to develop the smart home industry.

Microsoft Launches Home-Hub Smart Home Center

In early December 2016, Microsoft launched the Home-Hub smart home hub. It is actually a feature in Windows 10, primarily serving home users. The core service is combined with a PC equipped with the Cortana (Chinese name: Xiaona) assistant, targeting home users to provide intelligent integrated home services. This can provide users with calendar, spreadsheet, music, and many other functions as well as respond to queries for files and information.

Amazon's Red-hot Item Echo

At the end of 2014, Amazon launched the smart Bluetooth speaker Echo into the smart home industry field, combining with voice assistant Alexa to act as a smart home hub. After Echo receives the user's voice command, through Alexa one can control electrical home appliances, contact Uber, or make purchases on e-commerce platforms. For third-party manufacturers of smart home appliances, Amazon's open attitude can attract third-party smart home brands such as Vivint, idevice, Belkin, and Philips Hue to access their smart home products and systems. Echo's launch received widespread acclaim and quickly exploded into popularity.

Apple Announces the Apple HomeKit

Apple first released the HomeKit at the 2014 World Developers Conference (WWDC). This platform is one of the world's largest smart home ecosystems. At present, Apple controls various devices through the "Home" app installed on Siri, realizing the interconnection of various devices and enhancing the user's smart life experience.

From the perspective of China's situation, the present competitive landscape of the smart home market is gradually becoming clear. There are four competitive forces in the market: the first force is the traditional electrical home appliance manufacturer, represented by Midea, Haier, and other companies, carrying out smart transformations on original products and launching related smart home appliances and platform products; as with the example of Haier's U+ smart life open platform, these types of companies mainly rely on hardware revenue. The second force comprises of giant Internet companies like BAT deploying software, services, content, and other fields to strengthen cooperation with traditional home appliance manufacturers. For example, Tencent launched the Penguin Smart Community SaaS system, which has captured the smart home offline market and the smart community market. At the same time, it cooperated with companies such as ORVIBO to jointly outline smart living; Alibaba married Midea to utilize e-commerce and cloud service channels with the intention of changing the landscape of the smart home ecology. This force represents the general trend of cross-border integration, and business models are also developing in this direction of diversification. The third force comprises of outstanding hardware companies such as Huawei and Xiaomi, among others. In 2015, Huawei released its HiLink connection protocol to the smart home industry and absorbed Midea and China Telecom as its partners, with the core aim being to attain interoperability and solve the problem of fragmentation with smart homes. Such companies are usually well positioned, clearly aware of their strengths and weaknesses, and rationally lay out smart home–related products and services. The fourth force comprises of other companies such as operators and video websites. Operators mainly rely on the advantages of network operations to lay out relevant software, hardware, and smart application products; video websites mainly use smart television as a carrier, conducting operations by charging content service fees.

Haier Launches U-Home

Haier's U-Home is Haier's solution to smart home living. By using artificial intelligence as its technical backbone as well as semantic speech recognition, image recognition, clothing identification, and facial recognition as its entry points for interaction, it connects all home devices to the Internet through an information-sensing device and can interact with other electrical appliances in the home by means of making phone calls, sending text messages, accessing the Internet, and so on.

The Development Prospects of the Smart Home Industry

Optimizing and Improving Single Products, Expanding Application Scenarios

With the development trend of smart homes, market consumer groups have already formed a stable demand for smart home items. From the earliest control of Wi-Fi networks to today's fingerprint and speech recognition, interactive capabilities are also gradually rising, while the users of smart home products will also shift from early adopters to more ordinary consumers, covering a wider range of ages. For example, smart home security products received widespread attention in various exhibitions in 2016, and smart home systems such as smart lighting, control of electrical appliances, and others have also matured. As user demand steadily increases, product development will also flourish and diversify. Application scenarios will continue to expand in the areas of family healthcare and energy conservation, among others, on the basis of protecting family safety and improving the living environment.

Standards Are Tending Toward Unification, the Ecology Is Gradually Maturing

More and more manufacturers are becoming involved in the smart home industry and launching their own smart home ecosystems. However, current technical standards among enterprises have yet to be interoperable or shared. For example, wireless technology protocols that are well known to the public such as Wi-Fi, Bluetooth, radio frequency, and ZigBee each have their own advantages and disadvantages, and the respective alliances among manufacturers prevent smart home products from being interchangeable. Since respective manufacturers have already invested a lot and are unwilling to sacrifice their own interests, it is difficult to reach a consensus on the same platform. Product incompatibility affects user purchase choices and increases the cost of laying smart home systems. This is one of the bottlenecks in the development of smart home industry. At present, standards organizations, chip component manufacturers, operating system vendors, and voice interaction vendors are working hard toward interconnection. For example, Huawei's OpenLife Smart Home Solution proposes seven API standards and six integration frameworks to move toward cloud networking, center around the intelligent gateway, openly aggregate, and work with third-party vendors to improve API specifications and standards. It is foreseeable that in the future with the respective platforms of giant enterprises as the core, there will be a shift toward relatively unified behavioral norms or standards and a series of superior resources will be gathered. From single products to ecological circles, the intelligentization process will improve and accelerate and the smart home ecology will trend toward maturation.

The Issue of Smart Home Security Faces Challenges

The high-speed development of smart home products also underlies enormous security challenges. The application of the technology and construction of platforms are only one aspect; the security issues involved in network need to be solved urgently. With smart home products developing toward interoperability, any device that is infected by attacks will also affect other devices. According to 2016 Vormetric's poll of US smart home security, more than half of Americans worry that their smart home security and camera systems will be hacked, and 52 percent of respondents express similar concerns about Amazon Echo's smart home system being in danger of hacking. At China's "3.15" banquet in 2016, the problem of "hijacking" smart devices was already exposed. In turn, the importance of the security factor has led to the slow progress of connectivity. Foreign countries have already launched products to test smart home information security, while at present China still lacks specialized products targeting the information security protection and attack detection capability of smart home products; the access management system for smart home products has also not yet been fully established. Because there are many differences in the hardware and firmware of different smart home devices, it brings challenges to device security evaluation and feedback compensation mechanisms.

For the security challenges faced, the government and the industry should actively formulate and unify relevant safety standards, promote the establishment of a smart home security system, promote the safety of the industry from the source, and drive the development of the industry. Vendors should strengthen security and take appropriate protective measures to ensure the safety of devices and users' private data, encrypt and protect user data in devices, and set access rights to further improve data security. At present, users have yet to realize the importance of maintaining the safety of smart home equipment. In the future, they need to improve their security awareness and regularly inspect smart home devices plus the stored data inside them to ensure security.

The Interoperability and Interconnectedness of the Smart Home Is the Best Application Scenario for AI

Currently, most smart home products exist in the form of isolated single products. The scheme presents features of fragementedness as brands cannot be interconnected, hindering a good user experience. The development of artificial intelligence can completely evolve the control center of the smart home, connect the separate smart home items in the family, and form a complete smart home ecology. This will transform the weak linkage that was the smart home's former shortcoming while Wi-Fi, Bluetooth, ZigBee, and other networks can act as an interconnected foundation for the smart home. The interconnectedness of hardware communication standard, cloud network connection standards, and other core connections will be the general trend. The development direction of the entire smart home will be wholly compatible with data, providing users with a valuable smart family solution and enabling users to lead a smarter, more convenient living experience.

The Trend of Smart Home Development Will Continue to Improve

The development of machine learning, pattern recognition, and Internet of Things technologies has brought about various interactive modes to make home products more intelligent and user-friendly. Related products gradually developed from mobile phone control to human-machine interaction, which in turn were gradually replaced by other more optimized smart home system control models. Looking toward the future, artificial intelligence technology will transform smart homes from passive intelligence to active intelligence and even replace people in thinking, decision-making, and execution. Therefore, the "Artificial Intelligence + Smart Home" vertical is full of space for imagination, and as artificial intelligence technologies mature, the smart home industry is bound to open a new chapter.

Bibliography

Morgan Stanley Research. Robo-Advice: Fintechs Enabling Incumbent Win, February, 2017:34. (查不到,这是期刊上的文章还是书?查不到相关资料).

http://www.cnii.com.cn/technology/2015-11/26/content_1656183.htm. (网页无效, 需提供网上名章的名称和日期).

"2012—2020年中国智能家居市场发展趋势及投资机会分析报告". 9 August 2012. http://smarthome.qianjia.com/html/2012-08/10_125666.html.

"2016年我国智能家居市场规模达605.7亿元". 18 January 2017. http://d.qianzhan.com/xnews/detail/541/170118-2ab6cf16.html.

"2017国内外智能家居发展现状报告". 25 February 2017. http://smarthome.ofweek.com/2017-02/ART-91002-8420-30107968_2.html.

"美国成为全球智能家居市场容量最大的国家". 27 May 2016. http://news.qq.com/a/20160527/019714.htm.

CHAPTER 10

Unmanned Aerial Vehicles

Unmanned aerial vehicles, otherwise known as drones, are unmanned aircraft that uses radio control equipment and an independent program control device. This includes unmanned helicopters, fixed-wing aircraft, multi-rotor aircraft, unmanned airships, and parawings. A broad definition also includes adjacent space vehicles (20 to 100 km airspace) such as stratospheric airships, high-altitude balloons, and solar unmanned aerial vehicles. This chapter does not discuss unmanned aerial vehicles in the broad sense, but instead specifically refers to unmanned aircraft that are equipped with artificial intelligence and other information and communication technologies.

The Vacant State of International Unmanned Aerial Vehicles Development

According to the American Teal Group's forecast, the size of the global unmanned aerial vehicle market will increase from \$6.4 billion in 2015 to \$11.5 billion in 2024, and the total market size will exceed \$89.1 billion. By 2024, the global market share of civilian unmanned aerial vehicles will increase by 12 percent, reaching \$1.6 billion. According to the predictions of British think tank International Institute for Strategic Studies, the demand for military unmanned aerial vehicles will increase by three times on the current basis in the next ten years and the market will exceed \$100 billion. The gradually forming global civil unmanned aerial vehicle market is also about to rapidly develop.

Tencent Research Institute et al. (eds.), *Artificial Intelligence*,
https://doi.org/10.1007/978-981-15-6548-9_10

With the intelligentization of and deepening tactical research conducted on unmanned aerial vehicles, unmanned aerial vehicles are expected to become mainstream military aircraft in the future, even replacing manned military aircraft. Between 2015 and 2024, the market share of military unmanned aerial vehicle systems used for reconnaissance and combat will reach around $72.7 billion. Of these, $40.8 billion will be used for unmanned aerial vehicle production, $28 billion for testing designs and development, $2–4 billion for maintenance services, and $18.1 billion, $7.1 billion, and $15.6 billion for equipment, ground control stations, and payload production, respectively.

Civil unmanned aerial vehicles are mainly divided into two types: consumer-grade UAVs and professional-grade ones. Because civil unmanned aerial vehicles have the advantages of relatively low cost, no risk of casualties, strong survivability, good maneuverability, and being convenient to use, they have been used in aerial photography; geological and geomorphological mapping; forest fire prevention; seismic investigation and nuclear radiation detection; border patrol; emergency disaster response; farmland information monitoring; pipeline inspection; wildlife protection; scientific research; maritime reconnaissance; fish monitoring; environmental monitoring; atmospheric sampling; rainfall enhancement; resource exploration; anti-drug enforcement; anti-terrorism enforcement; police investigations; security surveillance; aerial photography for firefighting; relaying communications; urban planning; smart city construction, and many other fields.

The United States, Japan, and other countries or regions have taken unmanned aerial vehicles as their development focus, fully supporting research to form robust R&D capabilities so as to maintain the strong development momentum of the unmanned aerial vehicle industry in the coming period. Among Asian countries, India has increased its procurement budget for systems. In addition, by 2024 the United States, which has the largest number of unmanned aerial vehicles in the world, will spend $11.9 billion in purchasing unmanned aerial vehicle systems.

Different Countries Have Different Development Advantages

The United States has always dominated the international unmanned aerial vehicle market and is the largest consumer of unmanned aerial vehicles, accounting for 35 percent of the total. Europe is the world's second largest unmanned aerial vehicle market, accounting for 30 percent of the total. The rest is shared among Israel, Russia, China, and other countries. China's local turnover accounts for 15 percent of the world's total. Although China's domestic consumer-grade unmanned aerial vehicle market has yet to fully open, Chinese-made unmanned aerial vehicle products have already "flown" overseas. In the international unmanned aerial vehicle market, what is good for China is the unmanned aerial vehicle market in developing countries. Because unmanned aerial vehicles can provide easy implementation solutions in national defense security, domestic security, and national resource exploration, developing countries have a strong demand for unmanned aerial vehicles but they do not have the ability to develop and manufacture unmanned aerial vehicles products on their own, relying instead on imports. All types of unmanned aerial vehicles that are relatively good value for money in China are relatively competitive in the above markets. From a demand perspective, Asia will become the largest customer of unmanned aerial vehicles systems in the next decade with an estimated spending of $20.5 billion, accounting for 50 percent of the total unmanned aerial vehicles production market. From an export perspective, companies in North America, Europe, and Israel will continue to dominate the unmanned aerial vehicles market. In addition, the annual production of unmanned aerial vehicles in Asia will double, amounting to $2.9 billion. Among them, South Korea will become one of the major suppliers, having developed a series of unmanned aerial vehicles including unmanned combat aircraft.

In terms of patent protection, the global patent application for UAVs entered a period of rapid development after 2000. Looking at the patent statistics of priority patent countries, the United States, Japan, and China are in the top three. The United States has clear advantages in the field of unmanned aerial vehicles patented technology; Europe and Japan have also formed certain patent advantages. As a new force in the field of unmanned aerial vehicles, China shows a clear trend toward an increasing volume of patent applications.

Unmanned Aerial Vehicle Applications Are More Extensive

At present, the use of unmanned aerial vehicles is becoming increasingly common in various fields. For instance, crop data monitoring, forest fire prevention, atmospheric sampling, artificial rainfall, survey resources, express delivery, power line, surveying, and mapping are all professional-grade unmanned aerial vehicle applications. Among them, agriculture is the most promising application field for civil unmanned aerial vehicles. At present, more than 2300 unmanned aerial vehicles have been used to spray pesticides and fertilizers in rice planting areas in Asia, and in Japan 90 percent of work of this type is completed by unmanned aerial vehicles. In future, the agricultural unmanned aerial vehicle market will be even broader. Pipeline and transmission line inspections are also one of the potential applications for civilian unmanned aerial vehicles. The length of US oil and gas pipelines exceeds 64,370 kilometers. In order to conduct at least six inspections per year, manned aviation equipment fly 120 million hours per year. It is expected that China's unmanned aerial vehicles will begin to be applied to pipeline inspections in 2017–2018. By 2025, the number of unmanned aerial vehicle flight hours is expected to exceed six million hours.

The demand for civil aerial vehicles is even stronger. The main suppliers of aerial photography aircraft are the French company Parrot and the American company 3DRobotics (3DR). Since 2010, Parrot has introduced AR unmanned aerial vehicles to the market and has sold more than 500,000 units. In 2014, sales have tripled compared to 2013.

In addition to the civilian sector, countries around the world are also aware of the enormous application potential and broad application prospects of unmanned aerial vehicles in the military field and have given extensive attention and support to the development of the unmanned aerial vehicle industry. For example, the Japanese army has planned to purchase three RQ-4B Global Hawk unmanned aerial vehicles.

In the commercial sector, foreign consumer-grade unmanned aerial vehicles have four main uses: entertainment, non-commercial secondary development (personal research), commercial secondary development (land, agricultural measurement), and commercial aerial photography (advertising, film, and television).

In addition, the importance of the unmanned aerial vehicle airspace management platform is increasing. Recently, AirMap, an unmanned aerial vehicle airspace management platform provider, is developing a flight map and alarm notification platform for unmanned aerial vehicles. At present, the AirMap system has been imported into most major airports in North America and is used by about 80 percent of the world's unmanned aerial vehicles.

Spraying Drones Have Already Been Applied to Near-Mass-Scale

In recent years, the United States, Israel, and other unmanned aircraft powers have integrated communication, navigation, VR, AI, and other concepts and technologies on unmanned aerial vehicles. The large-scale application of unmanned aerial vehicles will be the first case in which robots are to appear in our lives. And the emergence of multi-rotor unmanned aerial vehicles has given the spraying drones industry a new boom. As of October 2016, there were more than 6000 registered UAV-related enterprises in China, including more than 300 companies producing agricultural unmanned aerial vehicles.

Unmanned Aerial Vehicles for International Humanitarian Relief

Internationally, the UNICEF and the Malawi Government are embarking on a collaboration to test whether unmanned aerial vehicles can provide faster and more effective assistance in the event of humanitarian disasters such as floods and droughts. The unmanned aerial vehicle test was conducted in April 2017 in the Humanitarian Unmanned Aerial Vehicle Testing Corridor, which is located outside the city of Lilongwe, the capital of Malawi. Although unmanned aerial vehicles have been tested for commercial transport in countries such as the United States and New Zealand, Malawi's test is considered the first time that unmanned aerial vehicles have been used for humanitarian assistance and development.

The development of the domestic unmanned aerial vehicle industry is advancing triumphantly.

At present, China's unmanned aerial vehicles are developing rapidly and has received much attention as an emerging industry. Since 2015, many entrepreneurs have entered the market. Whether it be DJI, Zerotech, or numerous technology giants including GOPRO, Tencent, Xiaomi, Amazon, and Google, these companies are all competing to enter the market. According to statistics, in 2015, the sales volume of unmanned aerial vehicles in China was about 90,000 and the sales volume of consumer unmanned aerial vehicles reached RMB2.33 billion. By 2020, the annual sales volume is expected to reach 650,000, displaying a growth spurt.

At the same time, unmanned aerial vehicles are an important part of the aviation industry chain. *Made in China 2025* lists aerospace equipment as one of the ten key development areas and one of the important development directions of aerospace equipment manufacturing would be to promote the development of unmanned aerial vehicles. The Chinese government's series of issued documents such as the *Outline of the National Medium and Long Term Science and Technology Development Program* and *Several Opinions of the State Council on Accelerating the Revitalization of Equipment Manufacturing Industry* have ushered in new development opportunities for China's unmanned aerial vehicle field. According to the classifications of industry life cycle theory, the unmanned aerial vehicle industry has entered the initial period of industry growth, showing a significant trend of rapid market demand growth, technology maturity, and market growth rate.

The Development of History of Unmanned Aerial Vehicles in China

The development history of unmanned aerial vehicles in China spans over 40 years. China has already established a relatively complete development system for unmanned aerial vehicles and is close to the international advanced technology standards in small-, medium-, and short-range and medium-altitude long-endurance unmanned aerial vehicles. In recent years, the application field of unmanned aerial vehicles has been expanding and shows great potential in the fields of resource exploration, ocean monitoring, aerial photography, and agroforestry surveillance.

Before 2010, the Chinese civil aircraft market was small and, slow to grow, mainly used in disaster rescue, ground mapping, and other markets. After 2011, China's consumer-grade unmanned aerial vehicle market rose rapidly, as exemplified by DJI. In recent years, unmanned aerial vehicles have been widely used in aerial photography, logistics, and other fields. Unmanned aerial vehicle delivery and the use of unmanned aerial vehicles for aerial photography on popular reality shows have drawn widespread attention to civilian unmanned aerial vehicles. China's consumer-grade unmanned aerial vehicles are at the forefront of the international market and have certain advantages in terms of market share, R&D capabilities, manufacturing capabilities, and application breadth and depth.

China's Unmanned Aerial Vehicle Development Trend

The Unmanned Aerial Vehicle Security Market Has Good Prospects. In segmenting the professional field, professional-grade unmanned aerial vehicles are mainly used in agriculture, electric power (circuit inspection), and policing. Domestic unmanned aerial vehicle enterprises often focus on one area for deep cultivation, creating differentiated advantages. For example, XAG mainly focuses on logistics unmanned aerial vehicles and agricultural unmanned aerial vehicles; Zerotech pays more attention to large- and medium-sized unmanned aerial vehicles and security monitoring unmanned aerial vehicles.

On the demand side, 90 percent of China's domestic unmanned aerial vehicle market demand comes from the military and the police while other demand directions take up only 10 percent. The demand is mainly for target drones and unmanned reconnaissance aircraft with electronic optical/infrared reconnaissance platforms. In order to ensure the rapid development of China's economy in the future and upgrade marine safety, border security, and the rapid survey and update of land information to an unprecedented height, the demand for unmanned aerial vehicles in these areas will continue to grow.

The Unmanned Aerial Vehicle Security Market Has Good Prospects

According to the statistics of China Aviation Industry Development Research Center, between July 2015 and June 2016, the security field in the unmanned aerial vehicle industry had a scale of about RMB180 million, accounting for 12 percent of the total market. Although the market size of the UAV industry has a relatively large correlation with population sizes, Beijing, Xinjiang, Tianjin, and Tibet are the national key security areas and the demand for unmanned aerial vehicles in these areas is even greater.

Domestic Enterprises Accelerate the Setup of Unmanned Aerial Vehicles in the Express Delivery Industry

SF Express is testing the distribution effect of unmanned aerial vehicles in the Pearl River Delta region, collecting flight data, and providing data support for the future construction of the overall operation and commissioning system. At present, 100,000 unmanned aerial vehicles purchased by SF Express have been put in place and about to be officially commercialized. Antwork and China Post cooperated to establish China's first unmanned aerial vehicle express line. Last year, JD.com announced that it would use unmanned aerial vehicles to distribute orders from rural areas.

An Open Ecosystem Is Starting to Take Shape

Civil unmanned aerial vehicle enterprises in China have mostly opened secondary development platforms. For example, in November 2014 DJI released the SDK (software development kit) development platform for products from the Elf unmanned aerial vehicle series, encouraging developers to use this toolkit and develop application software for their respective needs. The core flight control platform can also be sold separately to unmanned aerial vehicle developers or other vendors.

Large-Scale High-Quality Enterprises Lead Domestic Development

DJI represents the global heavyweight companies that are leading the Chinese unmanned aerial vehicle industry toward a more mature development direction, and thereby leading the development of the global industry for small-scale unmanned aerial vehicles. In the Chinese market, before the launch of DJI's unmanned aerial vehicles the price of civilian unmanned aerial vehicles was high. The minimum price of the DJI UFO unmanned aerial vehicle was only RMB5999, which lowered the price of unmanned aerial vehicles to below $1000. This was nearly half the price of foreign products of the same level, and the price of the three-axis handheld PTZ is about one-fifth of that of foreign products, successfully entering the general public's range of purchasing power. The products of DJI Company to a large extent have also changed the product model of the industry. In the past, most consumer-oriented unmanned aerial vehicles required assembly knowledge, and being limited to a few professional players, the market was not popularized for a long time. DJI realized a foolproof style of control for small-scale unmanned aerial vehicles.

Although the international giants have, one after another, staked out areas of the unmanned aerial vehicle industry in the international market, in the fast-growing market of civilian small unmanned aerial vehicles, Chinese enterprises have occupied an absolute dominant position in both technology and sales. The domestic small-scale unmanned aerial vehicle companies, as represented by DJI, Zerotech, Ehang, and PowerVision, have developed rapidly, and their scale far exceeds that of foreign enterprises.

Bibliography

http://tech.sina.com.cn/e/2014-01-14/09379095013.shtml.

"无人机未来国内外竞争力市场空间分析". 3 October 2016. http://www.81uav.cn/uav-news/201610/03/20108.html.

"无人机产业崛起对我国制造业转型有何启示?" 11 August 2016. http://www.gkzhan.com/news/detail/90652.html.

CHAPTER 11

Artificial Intelligence Enterprises

The AI Enterprise Ecosystem

Artificial intelligence has undergone over half a century of development. Since 2015, on the basis of continuous improvement of artificial intelligence research and application scenarios at home and abroad, China's artificial intelligence–related research has entered a stage of rapid development. In 2017, the first instance of artificial intelligence entering government reports implies that it has risen to the national strategic level. Artificial intelligence development plans and supporting policies will continue to be introduced and the artificial intelligence sector will lead the investment field. Undoubtedly, all of this will stimulate the development of artificial intelligence in many fields and activate entrepreneurial enthusiasm in artificial intelligence and related fields. The artificial intelligence field ushered in the entrepreneurial boom after 2011, reaching the peak of entrepreneurship in 2014 and 2015, and the current average age of enterprises is 3.2 years.

If 2016 were to be seen by everyone as the first year of artificial intelligence, 2017 is considered as the "first year of artificial intelligence application," which has also made Internet giants more eager to invest heavily in the field of artificial intelligence. Whether it's Facebook establishing an artificial intelligence R&D center, IBM setting up an artificial intelligence platform, or domestic giant companies BAT setting up their own research institutes, it is not hard to see that the wave of artificial intelligence has already risen to become the focus of future IT industry

Tencent Research Institute et al. (eds.), *Artificial Intelligence*,
https://doi.org/10.1007/978-981-15-6548-9_11

development. With the repeated mentions of artificial intelligence in national policies, the growth of investment in artificial intelligence, and especially the multi-faceted investment into artificial intelligence by giant technology companies, many artificial intelligence companies have sprung up. It is undeniable that this is the best era for the development of artificial intelligence, providing the best soil for development and the best entrepreneurial ecology. It can be expected that the future landscape of the industry is likely to be the case that "giant technology companies hold the positions for artificial intelligence platforms and entrances, while startups expand vertically into deep application fields." In recent years, China's entrepreneurial boom has not diminished, and major giant companies have deployed artificial intelligence. Both industry giants and startup projects have the data advantage and are bound to gain a global leading position in the field of artificial intelligence.

The Distribution of the Artificial Intelligence Startup Field

Although artificial intelligence is at an earlier stage entrepreneurship compared to Internet enterprises, the social discussion and effects sparked by artificial intelligence far exceed those that were once sparked by the Internet, and the entrepreneurial projects that have joined are also developing at explosive rates of growth. According to incomplete statistics, as of 2016, there were about 1000 Chinese artificial intelligence enterprises and about half of them had already received investment. Merely within the year of 2016, about 280 Chinese artificial intelligence enterprises received investment, and outstanding projects that have emerged from the market have begun to explore a business model based on artificial intelligence and gradually complete the transition from Internet+ to artificial intelligence+.

At present, computer vision, robotics, and natural language processing are the hottest areas of entrepreneurship. This inseparably corresponds to application scenario and data accumulation at the early stages of the industry, and the level of China's facial recognition technologies level puts it in a leading position.

With the improvement of technology in recent years and the number of artificial intelligence talents continuously increasing, the market scale of artificial intelligence in China is expected to continue to expand. More and more entrepreneurial projects also fall into the vertically segmented fields of artificial intelligence, moving closer to actual needs of use. And with the birth and growth of many vertical companies, artificial intelligence will appear in more industrial-grade and consumer-grade application products.

The Artificial Intelligence Industry Value Chain

Drawing support from technology breakthroughs in image recognition, speech recognition, and semantic understanding, a large number of artificial intelligence applications are beginning to commercialize. Artificial intelligence–related technologies can be classified into three categories according to the life cycle of data processing and application: basic artificial intelligence technology, technical artificial intelligence technology, and applied artificial intelligence technology.

The Classification of Chinese Artificial Intelligence Entrepreneurial Projects

The influx of capital has brought the artificial intelligence industry to land. In 2016, venture capital firms, private equity firms, and others invested rapidly in the field of artificial intelligence, which is expected to increase at the level of $6 billion to $9 billion, three times compared with 2013. Artificial intelligence entrepreneurial projects are most widely associated at the application level: robots, drones, smart homes, virtual personal assistants. Most Chinese startups usually possess original industry data accumulation and technical resource advantages in their respective fields. They can make breakthroughs in a small segment of the field, drill deep, and gain market share through the continuous improvement of technology. At the same time, based on the artificial intelligence technology itself, there are also quite a few business-to-business enterprises that provide core technical services as their products, bringing about the actual implementation of artificial intelligence in China.

Industry analysts believe artificial intelligence is currently making breakthroughs in many fields, but the core is still the algorithm application behind the data. In the beginning, artificial intelligence is applied in highly specialized, segmented fields, and with the abilities of human-machine interaction, the most widely applicable areas such as smart driving, pan-entertainment, smart consultants, and smart business transactions will be the easiest to break through.

In the field of smart transactions: Interaction with enterprises begins with customer service as part of enterprise-level service. If one takes solely customer service, customer service can only occupy a small part of the service but with the upgrade of technical capabilities and the improvement of human-machine interaction capabilities, smart customer services can be gradually upgraded to become the next generation of smart business transactions. Through the smart interaction portal one can connect more functions, complete the personalized communication pipeline from mechanized single sentences to contextual communication, leap from after-sales to pre-sales service, and the vast majority of enterprise-class intelligent services. According to analysts, from the traditional customer service system to the intelligent customer service system to the enterprise intelligent interaction comprises a three-level leap and these are the real needs of many Internet companies.

"In the whirlwind of artificial intelligence, we have cultivated more than ten years of technical work in the Internet industry so we can let the team run even faster!" founder and CEO of Zhuiyi Technology, Wu Yue, said. This former head of Tencent TEG business group search department personally experienced the rapid development of the Internet PC era to the mobile Internet era and various battles of technical upgrades. In 2016, Wu Yue and several former core Tencent technical directors and technical backbones collaborated to formally join the whirlwind of artificial intelligence. As a center of natural semantic understanding research, the artificial intelligence company combines cutting-edge technologies such as machine learning and cognitive computing; it focuses on natural language understanding, sentiment analysis, and response; and works to discover more possibilities of "artificial intelligence" as well as applying scientific research that "gives machines wisdom." Zhuiyi Technology has already launched the main human-machine intelligent interaction using the robot Yibot that is based on deep semantic understanding, reasoning, and dialogue.

Regarding the expectations of the Chinese market, Laiye's Wang Guanchun and Hu Yichuan unanimously agreed: "Although the US' artificial intelligence enterprises lead the world in underlying technologies such as artificial intelligence chip, software architecture, and general machine learning and deep learning algorithms, China has a greater opportunity in the field of consumer-oriented artificial intelligence-driven products." Laiye is a comprehensive personal assistant service platform mainly providing cloud-based personal assistant services in first-tier cities.

Adopting the manual+intelligence model and with the support of computer programs, through a natural and intimate style of interaction, it meets the user's requests such as taxi calls, coffee buys, general purchases, air tickets, train tickets, hotels, massages, errands, takeaway orders, reservations, cleaning, express delivery, movies, and so on. It also connects to third-party platforms to execute tasks offline and is committed to connecting people and local businesses through the most natural styles of interaction such as dialogue and referrals. Regarding how to think about the impact of artificial intelligence on future labor, founder and COO of Emotibot, Zhao Yuying, remarked: "We do smart customer service not to replace manual customer service, but to better cooperate with them, help enterprises to bear fewer labor costs and improve work efficiency." Zhao Yuying believes that human productivity will increase by 20–30 percent due to the participation of artificial intelligence. In the next three years, intelligent customer service can replace 80 percent of manual work.

The field of computer vision: The use of facial recognition is currently the most widespread. Face detection, identity verification, and other technologies have been applied to various fields, typically represented by facial recognition companies including Megvii and Tencent. Megvii uses deep learning and computer vision as its core technologies to continuously expand its advantages in visual recognition and deep learning. It has provided more than 15 billion data services, making it one of the larger intelligent data providers in China. Tencent's YouTu team focuses on image development, pattern recognition, machine learning, data mining, and other areas to carry out technical R&D and launch business.

In the specialization of facial recognition, YouTu is using a multi-machine, multi-card cluster training platform. The platform is a machine learning cluster independently developed by the YouTu engineering team. It has a training framework for functions such as cluster scheduling, storage, and management. It supports most network models as well as YouTu's special network models and introduces distributed computing into deep learning. Not only has the time for deep model training been greatly shortened, but it has also provided the ability to train ultra-deep neural networks. This has enabled YouTu's facial recognition to be crowned in the one-million-level "Facial Recognition Test."

Healthcare: In the field of healthcare, the application of artificial intelligence has been extremely extensive. McKinsey's artificial intelligence industry report is very optimistic about the future forecast of the medical industry, predicting that artificial intelligence can save 30–50 percent of healthcare productivity in the future and save $2–10 trillion globally in medicine and treatment costs. From the perspective of application scenarios in the medical field, it is mainly divided among a total of 11 fields including virtual assistants, medical imaging, drug research and development, nutrition, biotechnology, emergency room/hospital management, health management, mental health, wearable devices, risk management, and pathology. A typically representative enterprise of this would be Huiyi Huiying. With internationally leading technologies in cloud computing, big data, and artificial intelligence, Huiyi Huiying has built a digitized, mobile, and smart medical imaging and tumor radiotherapy platform. It has also constructed an intelligent image screening system, an anti-misdiagnosis system, and a diagnosis and treatment system aided by artificial intelligence for tumor, cardiovascular, acute abdominal, and other singular diseases. In the early stages it used medical imaging as an entry point to provide image cloud systems, image recognition, and intelligent diagnostic services. Nowadays, it collaborates with scientific research institutions, and through establishing models of human organs and deep neural network technology, it has realized a high degree of tumor recognition and is one of the first to apply automatic diagnosis of chest and brain nuclear magnetics in the actual operation process. In the future, Huiyi Huiying will release automatic detection of carotid stenosis, nuclear magnetic resonance imaging of the brain, esophageal cancer detection, and cerebral infarction.

The field of smart recruitment: The latest intelligent data matching plus continuous resume updating and crawling technologies bring the two sides of the recruitment experience into a new era. On the smart recruitment platform, employers and candidates input the desired job information and talent qualifications. Through smart recruitment platform's big data matching and semantic understanding technologies, accurate two-way matching is achieved.

The field of smart legal services: The most direct application is the smart legal assistant. Legal services is a professional service field. At present, artificial intelligence products have no perfect independent solutions for complex disputes and are still dominated by people. Many legal projects are explored through a single point of entry. For example, "contractors" accumulate data through contract tools to provide enterprises with legal solutions based on big data and artificial intelligence. Xiaofa Bo builds legal robots, provides legal support for professional law students and lawyers, and looks forward to using his own legal robots in the future to exhibit

AlphaGo-type cases where legal robots compete with humans and truly bring artificial intelligence to the real world.

The field of autonomous driving: It is currently an extremely hot application field. Through the HUD transparent projection screen placed on the instrument panel directly in front of the steering wheel to display information, entrepreneurial projects such as Carrobot can make calls, send and receive WeChat messages, listen to and pick songs, and perform other functions to allow users to concentrate on driving while at the same time safely guide navigation and meet entertainment and social needs to make driving safer and more carefree. Through using artificial intelligence to liberate manpower, UISEE, Momenta, TuSimple, and other companies can reduce traffic accident rates and more. I believe that in the future, autonomous driving will transform our travels to become safer and more intelligent.

The field of smart investment: Asset allocation is an important duty of investment consultants. In the field of smart investment consulting, artificial intelligence technology and big data analysis are combined with the investors' financial status, risk preferences, financial goals, and so on to provide tailored asset portfolio recommendations for investors through established data models and back-end algorithms. Taking the first-mover advantage in the stock market is a typical entrepreneurial project in the field of smart investment. It uses big data technology to analyze market data and then uses artificial intelligence algorithms to model data and predict the direction of the secondary market to provide smart investment services for investors. At present, the field of smart investment is looked upon favorably by investors. Third-party smart investment platforms such as Sparanoid, Blue Ocean Wealth Management, and Jimubox are continuously emerging on the Chinese market, along with smart investment platform developed by Internet companies represented by JD Finance, Qimingpian, and Hithink Flush Information Network.

The field of smart education: Taking SquirrelAI as an example, which takes self-developed and self-adaptive learning systems as well as teaching content from teachers, it fosters a dynamic learning community, highly interactive learning environment, and maximizes the learning effect through a combination of online and offline learning modes. SquirrelAI successfully developed China's first adaptive learning engine with complete independent intellectual property rights and advanced algorithms as the core, providing various educational institutions with adaptive learning solutions. In addition, the voice evaluation software developed by iFlytek and Qingrui Education can quickly evaluate pronunciations and point out where the pronunciation is inaccurate.

The exploration of other industries in the field of artificial intelligence has also gradually begun. In addition, encouraging the use of the Internet of Things (IoT) in traditional industries will help artificial intelligence generate more value. The Internet of Things enables connectivity between devices through sensors and networks, providing massive amounts of real-world data for artificial intelligence. Combined with the "Internet Plus" policy, the government can help create successful cases of IoT applications in key economic sectors and set an example for other industries.

With the Ability of Artificial Intelligence as the Foundations, Combining with Traditional Enterprises

As a technology receiving the attention of a new era, artificial intelligence is in essence a technological development that raises all industries and not just a single industry. Only when artificial intelligence technology is truly universally applied to traditional industries in China and not just the technology giants will its economic potential be fully demonstrated.

Enhancing the productivity level of all walks of life through artificial intelligence will create huge value. Artificial intelligence will intelligentize hardware products and the application value of the industry will gradually be reflected, such as in the aforementioned applications of healthcare, education, and finance. The entrepreneurial project of artificial intelligence is only in the early stages of development and the improvement of traditional enterprises is just beginning to emerge. With the industrialization of artificial intelligence and the guidance and support of China's industrial policies, the development of artificial intelligence will inevitably drive traditional enterprises to upgrade the strength of their technology and raise the level of the entire industrial field.

AI Driving the Intelligentization of Industry: All Major Platforms Are Laying Out AI Ecology

The year 2017 was dubbed the "key year" of AI. Deep learning, image recognition, speech recognition, and other technologies had continuous breakthroughs and artificial intelligence has already displayed strong developmental potential to transform the layout of the technology industry.

In order to more fully greet the era of intelligent business, with the entrepreneurial boom of artificial intelligence, Chinese industry giants are laying out artificial intelligence, providing technology, capabilities, manpower, flow of goods, and so on, and other entrepreneurs are supporting artificial intelligence in multiple forms to create an ecological layout of the artificial intelligence industry. Tencent created an AI Lab, a special technical team, that integrates the original open capabilities to support artificial intelligence entrepreneurial projects with deep exploration of artificial intelligence use scenarios, in order to cultivate an artificial intelligence ecology.

AI Lab Builds Tencent's Core Technical AI Team

On March 23, 2017, Dr. Zhang Wei, a top scientist in the field of artificial intelligence, assumed the role of Tencent AI Lab's (Tencent Artificial Intelligence Laboratory) director.

Dr. Zhang Wei became the first person in charge of Tencent's AI Lab, leading more than 50 scientists and more than 200 artificial intelligence application engineers to focus on the basic research of artificial intelligence. At the same time, due to Tencent's own business needs, Tencent AI Lab will also cooperate in the research and application of the four directions of content, social, gaming, and platform tool-type AI.

Dr. Zhang Wei is a special expert in the "Thousand Talents Program" of the Organization Department of the Communist Party of China. He holds a bachelor's degree in mathematics and computer science from Cornell University and a master's and doctorate in computer science from Stanford University. Prior to joining Tencent, Dr. Zhang was a professor at Rutgers University, a researcher at IBM Research Institute, a research fellow at Yahoo Research Institute, the vice-president of Baidu Research Institute, and the head of Baidu's Big Data Lab. He participated in and led the development of multiple machine learning algorithms and application systems.

Zhang Wei said: "The Internet is divided into first and second halves. The first half is the PC Internet and mobile Internet era. With the support of the demographic dividend, traffic dividend, and content dividend, this era has basically ended. What is the second half? Everyone thinks artificial intelligence is a very important direction." In the diversity of opportunities brought about by artificial intelligence, the core technology has become the commanding height of enterprise strategies.

Tencent also hopes to shift from a product-oriented company to a technology-oriented company by leveraging its accumulated technological advantages over the years.

AI Lab places great emphasis on underlying and fundamental technology research such as algorithmic capabilities that rely on machine learning. On machine learning, it further looks at how the machine looks—that is, computer vision; how the machine listens—that is, speech recognition; how the machine understands—which is natural language processing, including text and interaction. After this research is completed, it will then be applied to the business level so as to directly generate value for the company. Therefore, application is also something that the AI lab puts a lot of effort into doing.

Based on this, AI Lab's deep mining capabilities include (1) artificial intelligence decision-making ability, which relies on reinforcement learning with FineArt as an example; (2) artificial intelligence comprehension, which relies on cognitive science research; and (3) artificial intelligence creativity, which depends on generating models.

Apart from this, AI Lab will also do some non-business-related, exploratory, futuristic frontier AI exploration. In general, all basic research will eventually be centered around the main business service areas.

Tencent Opens Its Platform to Create an Open AI Strategy

Tencent's open strategy is based on open development. With the Internet ecology's flourishing as the target, it aims to help Internet entrepreneurs develop as well as promote industrial chain cooperation and win-win situations. With the opportunity of mass entrepreneurship and innovation, it will open up its core resources to entrepreneurs. Joint entrepreneurship can help pioneers configure new engines.

At present, entrepreneurship is no longer about fighting alone but about forming entrepreneurial alliances such as Tencent's open platform. The two sides exchange resources and interests to form a close entrepreneurial partnership. Makerspaces provide a three-in-one entrepreneurial resource service, which connects the Internet+O2O+offline resources for entrepreneurial projects and greatly enhances the success rate of entrepreneurship.

Based on the AI open strategy built by Tencent's open platform, with the artificial intelligence accelerator as the pioneer, Tencent is bringing together top technologies, professional talents, and industry resources.

Relying on Tencent AI Lab, Tencent Cloud, YouTu Labs, and the powerful AI technology capabilities as partners, it is forging to upgrade AI entrepreneurship projects. Through the resources of the Tencent brand, venture capital, and traffic advertising, it is finding more application scenarios for AI technology and products, and realizing the whole process from product creation to ignition.

Tencent AI+Accelerator centers around product+technology and will connect the project with technical parties internal and external of the company, resource partners, and industry partners. Acceleration of the project will unfold around five dimensions:

- Technology: Design the acceleration process with AI technology+scenario as its core; provide technical framework services and various methods of connection through an open interface and customized development.
- Instructor: The accelerator invites the company's technical experts, product experts, and external industry experts as mentors.
- Industrial resources: provide connections between the supply chain, hardware, and other industries; connect industry customer resources.
- Market: Provide TGPC and other exhibition stages, AI industry media tracking reports.
- Investment: Connect investment institutions, connect Tencent's investment opportunities.

Regarding the position of Tencent's AI accelerator, Deputy General Manager of the open platform, Wang Lan, said: The accelerator's position in the whole ecology is as a comprehensive structure. We have the top unicorn enterprises and we also see that there is a large part of the industry are medium enterprises. Actually medium enterprises are forever the most prosperous in this ecology. Perhaps it's not a standard unicorn such as Midea or Didi but it carries a very rich function to the whole ecology and they will be applicable in each vertical type class where there are many rich long-tailed or segmented scenarios. What we hope the entire accelerator can achieve is to find the first place in each sub-field or to help it become the first in a very specific sub-field. And for the entire AI open platform, we hope in future to achieve Tencent's own three pillars of AI or three capabilities, plus a more open platform for outsourcing, and the AI capabilities of our partners can be put on the open platform to let more people use it.

As Lin Songtao, vice president of Tencent, conveyed the belief: The Tencent AI Accelerator will be dedicated to supporting potential entrepreneurs or entrepreneurial projects. By opening Tencent's AI capabilities and the process of technology accumulation, it will lower the threshold of AI entrepreneurship and help AI entrepreneurs industrialize technology. Together with partners, it will promote the upgrade of "intelligence" in various occupations and industries.

Entrepreneurship projects accompanying the rise of artificial intelligence are still in their infancy. Combined with Tencent AI's scientific research strength and open source support, we are happy to look forward to the prosperity of many parties and build an AI industry ecology.

PART III

Strategy: A Detailed Look at National Strategies

In recent years, the world has launched a wave of artificial intelligence research and development. The United States, Japan, the United Kingdom and other world science and technology powers are all paying close attention and have strived to incorporate artificial intelligence into national strategy. They have introduced relevant tactics to strengthen the top-level design of artificial intelligence development and seize the strategic advantage. Since 2016, this trend has become clearer and major countries have placed artificial intelligence in an important position to enhance their strategic position. This section will take you through a detailed look of how the world's major countries are strategizing in the field of artificial intelligence.

CHAPTER 12

Top-Level Plans

One after another, the major powers in the world have introduced national strategies in the field of artificial intelligence, speeding up top-level planning and seizing the dominant position in the artificial intelligence era. The US White House released three government reports on artificial intelligence in succession and became the first country in the world to take the development of artificial intelligence up to the level of national strategy. At the same time, the strategic plan for artificial intelligence was seen as America's new Apollo lunar landing plan, with the hope that the United States could possess the same hegemony in the field of artificial intelligence as it did during the Internet era. The United Kingdom set its development goals for artificial intelligence through its 2020 national development strategy and issued a government report to accelerate the application of artificial intelligence technology within the British government. Not limited to just the United States and the United Kingdom, the European Union started early in 2014 to launch "SPARC," the world's largest civilian robotics R&D program. The Japanese government enacted "Japan's Robot Strategy: Vision, Strategy and Action Plan" in 2015, proclaiming that Japan wants to revolutionize artificial intelligence robots. This series of top-level plans ranges from autonomous vehicles to precision medicine to smart cities, with investments concentrated in innovation sectors enabling transformations in key national domains, in response to the challenges facing nations and the world. Simultaneously, this also means that since the current development of science and technology has artificial

Tencent Research Institute et al. (eds.), *Artificial Intelligence*,
https://doi.org/10.1007/978-981-15-6548-9_12

intelligence as the goal, the future of humanity can never be separated from artificial intelligence. If governments, industry, and the public work together to support the development of technology and pay close attention to its development potential and manage its risks, then artificial intelligence will become the main driver of economic growth and social progress.

A Vast World, Full of Promise

Under the current top-level plans in various countries, the application and promotion of artificial intelligence in key areas are generally accelerating. Currently, artificial intelligence is being deployed in relatively mature areas such as transportation, finance, and medical treatment, while at the same time research and development is being strengthened in the fundamental areas of artificial intelligence, such as human brain research.

Self-driving Cars

Wi-Fi-connected, self-driving vehicles can improve the safety of public roads, as sensing, computing, and data science breakthroughs enable commercial use of inter-vehicle communications and advanced autonomous technologies. According to the new version of the "Strategy for American Innovation" (hereinafter referred to as the new version of the "Strategy"), the 2016 budget requires the federal government to double its investment in autonomous vehicle R&D. The "Roadmap for US Robotics 2016" raises the point that a new generation of autonomous driving systems has been used in cars, in aircraft, underwater, and in space probes. Since Japan, Germany, the United States, and South Korea are all major automobile manufacturers, the major demand for industrial robots in the future will still be in the automobile industry.

Precision Medicine

The new version of the Strategy proposes a Precision Medicine Initiative that will take advantage of genomics development, innovative approaches to managing and analyzing large data sets, and health information technologies, all the while paying attention to protecting privacy.

Advanced Manufacturing

The surge in demand for customization in manufacturing poses new requirements for artificial intelligence in this area. For example, in the automotive industry, a high-end car can have countless different configuration options, including everything from seat color to electronics configuration. The manufacturer's product line needs increasingly sophisticated technology in order to meet these demands.

Smart Cities

Smart cities are cities equipped with tools to address the most pressing issues of public concern such as traffic congestion, crime, sustainability, and the provision of essential urban services. In September 2015, the US government announced a new smart city project that invested over $160 million to help communities address key challenges such as reducing traffic congestion, fighting crime, boosting economic growth, coping with the effects of climate change, and advancing the provision of public services.

Human Brain Research

Understanding human brain functioning is not only crucial for the healing of brain-related diseases, but also revolutionary for the development of computers that resemble human brains. The human brain plan released by the EU and the US BRAIN plan have given rise to a new wave of global brain science research. Japan has also released a corresponding plan for brain science research.

World Powers "Devise Their Battle Plans

"Defend the Lead" America: A Comprehensive, Strategic Positioning

In October 2016, the White House released the "National Artificial Intelligence Research and Development Strategic Plan," which became the world's first strategic plan for the development of AI at the national level. The program aims to use federal funds to continue to deepen understanding and research of AI, so that this technology makes a more positive contribution to society, with negative impacts reduced. This US plan has major significance as a plan for all countries, especially China, to refer to and borrow from when formulating future AI development strategies. The plan states that the United States will make sustained investments in a number of areas: advancing data-focused methodologies for knowledge discovery; enhancing the perceptual capabilities of AI systems; understanding theoretical capabilities and limitations of AI; pursuing research on general-purpose artificial intelligence; developing scalable AI systems, fostering research on human-like AI; developing more capable and reliable robots; advancing hardware for improved AI; and creating AI for improved hardware.

In October 2016, the Executive Office of the President and the National Science and Technology Commission jointly released the report "Preparing for the Future of Artificial Intelligence." In order to help the US government to cope with developing trends in artificial intelligence, this report analyzes the status quo, existing and potential applications of artificial intelligence, and related social and public policy issues. The US government will provide substantial investment to help research artificial intelligence and the government has decided to become an early client for the technology and its applications. The US government believes the outlook is extremely optimistic for artificial intelligence and machine learning and that they will enable people to lead better lives. The report's assessment shows that long-term concerns about super-IQ, strong artificial intelligence have almost no impact on current policies. At the same time, it points out that due to the rapid development of artificial intelligence, there is also a huge demand for support from personnel with relevant skills and for the development of related fields. All citizens need to prepare to receive education on AI. In addition, it is particularly important to prevent machines from causing prejudices and to ensure the "morality" of artificial intelligence, so as to ensure that artificial intelligence can promote justice and fairness and artificial intelligence–based technology can obtain the trust of stakeholders. The report points out that ethics education for AI practitioners and students is also an important part of the equation.

On October 31, 2016, more than 150 research experts from the United States jointly completed the 2016 edition of "A Roadmap for US Robotics, From Internet to Robotics." As a roadmap for the development of national robots, it introduces topics such as: the transformation of manufacturing and supply chain; next generation consumer and professional services; healthcare; enhancing public safety; Earth and beyond; workforce development; shared infrastructure; and legal, ethical, and economic issues.[1] The roadmap calls for a better policy framework to safely incorporate new technologies such as autonomous vehicles and commercial UAVs into everyday life. It encourages increased research in the field of human-computer interaction, so that people can live in their own homes in their old age. It calls for an increase in STEM-related educational content from primary school to adulthood. It calls for research to create more flexible robotic systems to accommodate the growing customization needs of manufacturing, from cars to consumer electronics. It proposes related recommendations to ensure continued leadership in robotics, whether in terms of research innovation or in technology and policy, and to ensure that research can be put into practice and really solve real-life problems. In terms of specific content, the roadmap synthesizes a wide range of applications of robotics, including driverless cars and their policies, healthcare and companion robots, manufacturing, industrial Internet, and the Internet of Things, education, shared robotics infrastructure, and legal, ethical, and economic issues. Lastly, the roadmap proposes that government at all levels should continue to accrue expertise in cyber-physical systems, foster innovation in robotics, maximize its potential for social good, and minimize the potential for harm. At the same time, it is necessary to support interdisciplinary research conducted by government and academia. Government and academia should actively cooperate to break down silos between disciplines and eliminate research obstacles. Independent researchers should be assured that there is no risk of violating existing laws and principles.

At the end of October 2015, the United States released a new version of its "Strategy," highlighting nine areas of focus related to artificial intelligence. The innovation strategy was first released in 2009 to guide the work of federal agencies to ensure that the United States continues to lead the global innovation economy, to develop future industries, and to help overcome various difficulties encountered in economic and social development. From the 2007 "America Competes Act" to the 2009 "American Recovery and Reinvestment Act" and "A Strategy for American Innovation: Driving Towards Sustainable Growth and Quality Jobs," and then to the 2011 "Strategy for American Innovation: Securing our Economic Growth

[1] It appears the author took the headings of the table of contents for the roadmap. The roadmap can be found here: http://jacobsschool.ucsd.edu/contextualrobotics/docs/rm3-final-rs.pdf.

and Prosperity," the United States has always attached great importance to the design of innovation strategies. The new "Strategy" follows the 2011 policy of sustaining the US innovation ecosystem and for the first time presents six key elements for sustaining such an ecosystem. The ingredients for innovation include the Federal Government investing in the building blocks of innovation, promoting private sector innovation, and empowering a nation of innovators. Building off these are three strategic initiatives that focus on creating quality jobs and sustained economic growth, catalyzing breakthroughs for national priorities, and providing the American people with innovative government. Based on this, the new Strategy highlights nine strategic areas: advanced manufacturing, precision medicine, BRAIN Initiative, advanced vehicles, smart cities, clean energy and energy-efficient technologies, educational technology, space exploration, and new frontiers in computing. The new version of the Strategy is the best source for understanding future strategic investment in the United States. Key areas mentioned, such as self-driving cars, smart cities, and digital education, are closely related to artificial intelligence.

As early as 2013, the United States launched the "Brain Research through Advancing Innovative Neurotechnologies" (BRAIN) initiative, which covers brain dynamics, new brain technologies, and interdisciplinary (physics, biology, sociology, and behavioral sciences) brain research. Since the program was promulgated, tens of top high-tech enterprises, university research institutes, and scientists have responded to the plan and successfully promoted its implementation. First, public and private agencies worked together to promote the BRAIN program. On the public agency side, the US government has been steadily increasing the budget for this project in recent years. The National Institutes of Health (NIH) is devoted to developing and applying new tools to map out the circuits of the brain. The Department of Defense Advanced Research Projects Agency (DARPA) facilitates data processing, imaging, and advanced analytical techniques. The National Science Foundation (NSF) develops physical and conceptual tools needed for the brain functions of a variety of living things including humans. In the private sector, the Allen Institute for Brain Science researches brain activity in cognitive, decision-making, and command operations; the Howard Hughes Medical Institute focuses on the development of imaging techniques and on neural network information storage and processing; the Kavli Foundation is devoted to studying mechanisms of brain disease and looking for treatments; the research of the Salk Institute for Biological Studies, which links single genes to neural circuits and then to behavior, reveals an in-depth understanding of the brain. Overall, private and public agencies are evenly matched in terms of their investment in the BRAIN program.

Second, they carry out cooperation between international projects to avoid duplication of efforts. In March 2014, they worked with the EU's Human Brain Project (HBP) to try to cover as many areas as possible under the premise of not duplicating work. Although there are more uncontrollable factors between the current state of brain science research and the expected accomplishments—even the technology to achieve the desired goal is still under development—it is imperative for the United States to promote this program. While moving forward, it should constantly clarify the specific details of achieving the goal and improve management. As the plan progresses, the White House will set up a coordination group for the agencies involved, helping them work together to achieve the goal.

In sum, the United States is, at this point, the country that has introduced the most strategies and policy reports on artificial intelligence. The United States is undoubtedly the forerunner in the field of artificial intelligence research and its every move necessarily affects the fate of all of humanity.

Ambitious EU—"Human Brain" and "SPARC" Projects

The EU also plans to promote artificial intelligence, focusing on human brain research and robot development. In 2013, the European Union proposed a ten-year Human Brain Project, currently the most important human brain research project in the world, which will simulate the brain through computer technology and establish a completely new and revolutionary information and communication technology platform for generating, analyzing, integrating, and simulating data so as to facilitate the applicability of the research results.

In addition, the European Commission, in collaboration with euRobotics,[2] finalized the "SPARC" program as the world's largest pri-

[2] euRobotics, an industry association headquartered in Brussels, Belgium, was established on September 17, 2012, by 35 institutions and now represents more than 250 companies, universities, and research institutes, ranging from traditional industrial robot manufacturers to makers of agricultural machinery and innovative hospitals, with very strong scientific and technological strength.

vately funded robotics innovation program to maintain and expand Europe's leadership and ensure Europe's economic and social influence. In terms of operational models, the program adopts a public-private partnership (PPP) approach. On the private side, a panel of expert members from the private euRobotics AISBL (French acronym for nonprofit international association) works through "task forces" to provide a high-level strategic overview through the Strategic Research Agenda (SRA), updating the document based on market and industry conditions, disseminating the ideas and intentions of private parties. The documents attached to the SRA also include a more detailed technical guide, the Multi-Annual Roadmap (MAR), to identify expected progress within the community and provide a detailed analysis of medium-term research and innovation goals.

Robot Superpower Japan—"New Industrial Revolution"

Japan's robot industry accounts for a much larger proportion of its national economic growth than in other countries. For the past 30 years, Japan has been called the "robot superpower" and has the world's largest number of robot users, robotics equipment, and robotics service providers and manufacturers. In recent years, with the decline of the Japanese birth rate, the aging population, and the shrinking of the population of childbearing age, among other increasingly serious social issues, robot technology has received more attention. Based on these problems, the Japanese government revised the "Japan Revitalization Strategy" adopted by the Cabinet in June 2014 and proposed that "a new industrial revolution driven by robots" (hereinafter referred to as "robot revolution") should be promoted. In order to achieve the goal of robot revolution, the Japanese government established the "Robot Revolution Realization Council" (hereinafter referred to as "the council") in September 2014. The council consists of a large number of experts with rich professional expertise. The council sessions focus on specific initiatives such as technological advances related to the robot revolution, regulatory reforms, and global standards for robotics. Japan's Ministry of Economy, Trade and Industry summarized the results of the committee's discussions and prepared the report "Japan's Robot Strategy: Vision, Strategy and Action Plan (hereafter referred to as the "Strategy"), which was released in January 2015. The specific content of the Strategy is divided into two parts. The first part is an overview and is divided into two chapters: The first chapter introduces the background of the development of the robot industry in the international community and the goal of the robot revolution in Japan; the second chapter introduces three key strategies to realize the robot revolution. The second part is the "five-year plan" for the development of Japanese robots, which is divided into two chapters. The first chapter elaborates on eight cross-cutting issues, including the establishment of a "robot revolution incentive mechanism," technological development, robot international standards, field testing of robots, and so on; the second chapter explains the development of robots in specific areas, including manufacturing, services, healthcare, and nursing.

Through research and development and popularizing robot technologies, the Japanese government hopes to ease the labor shortage problem and liberate humankind from overwork, as well as increase the productivity of manufacturing, medical services, and nursing, as well as agriculture, construction, infrastructure maintenance, and other industries.

Unwilling to Fall Behind Britain—Facing the Fourth Industrial Revolution Challenge

The British government is preparing for Brexit and currently embarking on the task of building and consolidating its own unique science and technology regulatory system and focusing on the development, deployment, and use of artificial intelligence systems and robots. This industry is crucial to Britain strengthening its leadership in the global socio-economic, technological, and intellectual fields and is consistent with the British government's industrial development strategy. The United Kingdom selected "Robotics and Autonomous Systems (RAS)" as part of its "Eight Great Technologies" program and announced that Britain wants to be a global leader in the fourth industrial revolution. In 2013, with the support of the Innovate UK program, a "special interest group" of academic researchers and industry representatives was formed to enhance cooperation and innovation in robotics and autonomous systems. In a RAS 2020 National Strategy released in July 2014, the group set out the development goal for RAS in the United Kingdom, that is to "capture value in a cross-sector UK RAS innovation pipeline through co-ordinated development of assets, challenges, clusters and skills." Eight recommendations were made to that end. Those worthy of attention include establishing a centralized leadership system to guide and supervise innovation activities; promoting international cooperation among governments, industry sectors, and agencies in order to encourage further innovations and strengthen ties; disclosing more information to the public; and institutionalizing Britain's position as a good investment market for global technological innovation and development.

In October 2016, the Science and Technology Committee of the UK House of Commons released a report on AI and robotics. The United Kingdom considers itself to be a global leader in ethical standards for robotics and AI systems. At the same time, its leadership in this area should extend to the field of artificial intelligence regulation. The report brings together a diverse range of experts and practitioners in robotics and artificial intelligence systems to explore the development and application of automated systems with advanced learning capabilities and the unique set of ethical, practical, and regulatory challenges that they bring. The report calls for the involvement of government regulators and the establishment of a guiding organization to ensure that these advanced technologies can be integrated into society and be beneficial to the economy. The report expounds on the potential ethical and legal challenges posed by the innovative development of artificial intelligence and its regulation, and attempts to find solutions that can maximize the socio-economic benefits of these scientific advances while minimizing their potential threats. The report stresses that the solutions outlined are of crucial importance in establishing and maintaining public trust in the government as AI technologies become more integrated into society.

In addition to the artificial intelligence development strategies, plans, and other documents released by the world's major powers, relevant international organizations are also increasingly concerned about the development of artificial intelligence. For example, the United Nations "Report on Robotics Ethics" provides a look at the new path of artificial intelligence systems based on the physical form of robots, serving as an effective complement to the perspectives of "national centers" in various countries in the world. The "Preliminary Draft Report on Robotics Ethics" (2015) jointly published by UNESCO and the World Commission on the Ethics of Scientific Knowledge and Technology focuses on the advancement of artificial intelligence through the manufacture and use of robots, as well as the resulting social and ethical issues brought on by these advances.

China, from "Running After" to "Setting the Pace"

Since the reform and opening up period, China has always attached a great degree of importance to the development of science and technology, sincerely believing that science and technology are the primary productive forces. In today's new economic normal, there is a greater need for a new revolution in science and technology to promote the transformation and upgrading of the economic structure and the long-term healthy development of the national economy. Artificial intelligence technology undoubtedly represents the highest level of science and technology today. On July 20, 2017, the State Council formally issued the "New Generation Artificial Intelligence Development Plan" (hereinafter referred to as the "Plan") and elaborated the Plan for the development of artificial intelligence in China from all aspects including strategic situation, overall requirements, resource allocation, legislation, and organization. The Plan points out that the overall development level of artificial intelligence in China still lags behind that of developed countries in terms of major original achievements, basic theories, core algorithms, key equipment, high-end chips and components, and so on. It proposes a three-step development strategy target for 2030: By 2020, the overall technology and application of artificial intelligence in China will be in line with the world's advanced level. By 2025, there will be major breakthroughs realized in basic theory. By 2030, artificial intelligence theory, technology, and application will all reach world-leading levels and China will become the world's leading artificial intelligence innovation center. Prior to the issuing of the Plan, the State Council formulated and released the "13th Five-Year Plan for National Science and Technology Innovation," the "13th Five-Year Plan for Developing National Strategic and Emerging Industries," as well as the "'Internet +' and AI Three-year Implementation Plan" jointly issued by the NDRC and various departments. All take the development of artificial intelligence to be a strategic priority, but it had not yet risen to the national strategic level. Overall, the Plan conforms to the wave of artificial intelligence development, and more comprehensively expounds the key issues in the development of artificial intelligence. It is the top-level plan for China's industrial wave.

Compared with the national strategies of other countries, the planning of China places more emphasis on technology and application, and comparatively plays down other aspects or problems of artificial intelligence development, such as human capital and education, standards, and the data environment. In the United States, for example, in October 2016, the White House released two important reports: "National Artificial Intelligence Research and Development Strategic Plan" and "Preparing for the Future of Artificial Intelligence." The "National Artificial Intelligence Research and Development Strategic Plan" serves as the world's first national-level strategic plan for artificial intelligence development (which Obama calls the new Apollo Lunar Program). It proposes seven strategic directions for the development of artificial intelligence, including: basic research strategy; human-computer interaction strategy; sociological strategy; security strategy; data and environment strategy; standards strategy; and human resources strategy. These seven strategic directions run in parallel, with the "National Artificial Intelligence Research and Development Strategic Plan" explaining each in depth. "Preparing for the Future of Artificial Intelligence" elaborates on steps to prepare for and guarantee the development of artificial intelligence in the areas of policy formulation, government regulation of technology, financial support, universal education on artificial intelligence, and prevention of machine prejudice. At the same time, it proposes 23 actions for implementing artificial intelligence. China, on the other hand, plans to put forward six key tasks, including:

1. establishing an open and collaborative artificial intelligence system for scientific and technological innovation;
2. fostering a high-end and efficient smart economy;
3. building a safe and convenient smart society;
4. strengthening military and civilian integration in the field of artificial intelligence;
5. building a safe and efficient smart infrastructure system; and
6. forward-looking planning of a new generation of artificial intelligence science and technology mega-projects.

All of these are in the area of technology or application, and there are a few proposals for investment, education, personnel, ethics, system construction, and other aspects.

In the wave of innovation, institution building is a form of productivity. The success of Silicon Valley in the Internet age was due in large part to the major US reforms to copyright and tort law in the 1990s, which reduced the responsibility of Internet platforms and provided a conducive legal environment for the success of Silicon Valley businesses in the Web

2.0 era. Only then, talented programmers were able to bring to bear their wisdom and intelligence, bringing forth stunning product innovations. In the AI era, the importance of institution building cannot be ignored. Taking data liberalization as an example, current artificial intelligence is based on the "feeding" of big data. Without data liberalization policy from government, many AI applications will become "water without a source, a tree without roots." It can be said that the issue of data liberalization is a pain point in the development of AI in China and needs to be elaborated upon in a more comprehensive and in-depth manner in the strategy. In addition, artificial intelligence–related legislation and supporting issues are also worth exploring in depth. The current plan only makes simple references to these. Take autonomous driving, a relatively mature technology in the field of artificial intelligence, as an example. In September 2016, the US Department of Transportation released the "Federal Automated Vehicles Policy" to provide a guiding regulatory framework for the safety testing and application of autonomous driving technology, laying out the direction for industrial development. At present, China still lacks such legislation and standards, which will have a significant impact on the development of the industry. Such issues are also worth studying in the planning process.

Recently, the *Economist*[3] pointed out that five major factors have pushed China to develop into a global AI center: (1) multiple industries wanting to use AI to achieve digital transformation; (2) a large number of highly talented people in artificial intelligence; (3) a promising mobile Internet market; (4) high-performance computing technology; and (5) government policy support. Among them, the first two factors are particularly important and are the unique advantages for China in becoming a global AI center. In the area of information and communications, in terms of broadband deployment, big data, and cloud computing, China is basically a strategic follower. In terms of AI, China followed the United States and Canada in releasing a national AI strategy. In the wave of AI industry, China should go from system follower and move toward being a leader, actively seizing the strategic high ground. Taking AI ethics as an example, overseas the "Asilomar" principles have been put forward for the development of artificial intelligence. The IEEE, UN, and others have promul-

[3] I believe the author is referencing this article: https://www.economist.com/news/business/21725018-its-deep-pool-data-may-let-it-lead-artificial-intelligence-china-may-match-or-beat-america.

gated relevant ethical principles of AI such as the principle of protecting human interests and basic rights, the principle of security, the principle of transparency, and the principle of being beneficial and inclusive. These are also emphasized in various countries' national strategies. China should also actively construct guidelines for artificial intelligence ethics and play a leading role in promoting the inclusive and beneficial development of artificial intelligence. In addition, China should actively explore ways to go from being a follower to being a leader in areas such as AI legislation and regulation, education and personnel training, and responding to problems with AI.

Bibliography

Brain Research through Advancing Innovative Neurotechnologies.

Department for Business, Innovation & Skills and The Rt Hon David Willetts. "Eight great technologies". 24 January 2013. https://www.gov.uk/government/speeches/eight-great-technologies.

CHAPTER 13

The Power of Capital

Artificial intelligence entered a stage of explosive growth from 2011. As of 2016, there were more than 1000 artificial intelligence companies worldwide. According to a report released by PricewaterhouseCoopers at the 2017 Summer Davos forum, artificial intelligence is expected to boost world economic growth by 14 percent by 2030, equivalent to $15.7 trillion, a figure that is higher than the combined size of the Chinese and Indian economies. According to the "State of Artificial Intelligence" report released by CB Insights in August 2017, in the preceding five years, the total amount of artificial intelligence startup financing exceeded $14.9 billion, and the total volume of transactions reached 2250. The rapid development of the artificial intelligence industry and technology has brought about a wave of financing in this field. According to statistics, "venture capital funding in the artificial intelligence sector grew from $3.2 billion in 2014 to $9.5 billion in the first five months of 2017. Since 2015, the number of investments almost doubled. The prominent business consulting company Frost & Sullivan listed artificial intelligence as the hottest investment area in 2017." In addition to the enthusiasm of the financial markets, governments have also increased policy and capital support for artificial intelligence. They have not only used national strategies to call for increased investment in artificial intelligence but also invested heavily in key research and development projects. The flow of capital toward the artificial intelligence sector is an increasingly prominent trend across the world.

Tencent Research Institute et al. (eds.), *Artificial Intelligence*,
https://doi.org/10.1007/978-981-15-6548-9_13

Funding Is the Foundational Guarantee for the Vigorous Development of Artificial Intelligence

Capital is the material guarantee for the development of artificial intelligence, and is a necessary condition for realizing foundational research and development and the rapid development of the industry. Only with sufficient financial security can it be possible to achieve breakthroughs in artificial intelligence research and development and continuously increase the market share of the industry. At present, the countries at the forefront of artificial intelligence development have all invested massively in the sector. The United States has become the current leader in the field of artificial intelligence research, not only because of its pre-existing strengths in science and technology but also because of the government's huge investment in research and development. In 2015 alone, the US government invested up to $1.1 billion in research and development related to artificial intelligence. The Korean government invests up to $100 million a year in research and development of artificial intelligence and robotics. The Japanese government provided $350 million in one year for research into intelligent assistant robots alone. At present, with the field of artificial intelligence in a stage of vigorous development, it is even more necessary for countries to grasp the momentum of development, increase the scale of financial support, and introduce policies to encourage investment in artificial intelligence. Only in this way will it be possible to seize the initiative in the wave of the fourth industrial revolution.

Money is an effective way to attract talent. Although generous material compensation is not the only condition for attracting talent, it is an extremely effective path. Through the improvement of remuneration packages for artificial intelligence talents, we can allow talents to be free of worries and ensure the smooth progress of research and development and management. At the same time, adequate financial security is also conducive to building the confidence of relevant professionals in the development of the artificial intelligence sector and increasing their commitment to professional work. The dual guarantee of capital + manpower is the key to getting the upper hand in artificial intelligence. In July 2017, LinkedIn, the world's largest workplace social platform, released the industry's first Global AI Talent Report. According to the report, the number of AI jobs published through the LinkedIn platform alone grew from 50,000 in 2014 to 440,000 in 2016, almost a nine-fold increase. In terms of the specific field, demand is currently strongest at the AI foundations level, especially in algorithms, machine learning, GPUs, smart chips, and so on. The talent gap here is more significant than at the technical level and the application level. In the context of the scarcity of professional talents, technology giants have increased their bargaining chips for attracting talent. Facebook's strategy to attract talent includes providing salaries of hundreds of thousands of dollars and work locations around the world.

Governments Increase Investment in Artificial Intelligence

At present, major countries have already targeted strategic opportunities in artificial intelligence development, and are successively introducing artificial intelligence strategies at the national level, launching key artificial intelligence projects and increasing financial support.

The United States

The United States has always focused on artificial intelligence research and development, and this has accelerated in recent years. As early as the 2013 fiscal year, the US government invested $2.2 billion of the national budget in advanced manufacturing. The National Robotics Initiative was one of the key areas of investment. In April 2013, the US government launched the Brain Research through Advancing Innovative Neurotechnologies program, which plans to invest $4.5 billion in ten years. In May 2016, the White House established a subcommittee on machine learning and artificial intelligence to coordinate actions relating to artificial intelligence across sectors and to explore related policies and laws. In October of the same year, the Executive Office of President Obama issued *Preparing for the Future of Artificial Intelligence* and the *National Artificial Intelligence Research and Development Strategic Plan* to raise artificial intelligence to the national strategic level of the United States. Two months later, the White House released the report *Artificial Intelligence, Automation and the Economy.* This discussed the expected impact of artificial intelligence–driven automation on the economy and described a wide-ranging strategy to increase the benefits of artificial intelligence and reduce its costs.

Preparing for the Future of Artificial Intelligence noted that according to public data, in 2015, the US government invested about 1.1 billion US dollars in research and development related to artificial intelligence. In all the artificial intelligence–related seminars and public promotion events hosted by the White House Office of Science and Technology Policy, industry leaders, technologists, and economists all called on government officials to increase government investment in artificial intelligence technology research and development. The Economic Advisory Committee analyzed that not only in the field of artificial intelligence research and development, but in all scientific research fields, tripling or quadrupling R&D investment would bring a net benefit in economic growth and would be a worthwhile investment for a country. To be sure, the private sector will be the main engine for the development of artificial intelligence technology. However, looking at the status quo, investment in basic research is far from enough. For basic research, which has the pure purpose of pushing the scientific boundaries in the field, the investment period is long. Therefore, it is difficult for private enterprises to obtain corresponding investment returns in the short term. *Preparing for the Future of Artificial Intelligence* suggests that the federal government should give priority to basic artificial intelligence research and long-term research projects. If the federal government and private companies can make stable, long-term investments in artificial intelligence research and development, especially in the high-risk basic research field, the whole country will benefit.

European Union

At present, the number of industrial robots worldwide is growing at a rate of 8 percent. Europe's share of the global market is about 32 percent, while its share of the global market for service robots is 63 percent. In order to maintain and expand Europe's leadership and ensure its economic and social influence, the European Commission and the European Robotics Association (euRobotics) collaborated to launch the "SPARC" program. The European Commission is funding SPARC under the Horizon 2020 program. According to the agreement, the European Commission invested €700 million[1] and the European Robotics Association has contributed €2.1 billion, making SPARC the world's

[1] In order to restore the economic strength of EU member states, the European Commission formulated the Europe 2020 strategy, which proposes three strategic priorities, five quantifiable targets, and seven supporting flagship initiatives. Horizon 2020 was

largest privately funded robot innovation program. In 2013, the European Union proposed the Human Brain Project, which will last for ten years. The EU and participating countries will provide nearly €1.2 billion in funding, making it the most important human brain research project in the world.

Global Giants Have Successively Joined the Artificial Intelligence Camp

In recent years, artificial intelligence technology has continuously made breakthroughs, and its application in fields such as finance, medicine, and manufacturing has expanded rapidly. McKinsey & Company expects the market for artificial intelligence applications to reach a total value of $127 billion by 2025. Artificial intelligence has become an area in which many governments and companies are competing. According to statistics, artificial intelligence firms are currently concentrated in a small number of countries, with the number of enterprises in the United States, China, and the United Kingdom accounting for 65.73 percent of the global total. Among them, there are more than 2900 artificial intelligence companies in the United States, more than 700 in China, and more than 360 in the United Kingdom. In the seven years from 2010 to 2016, Chinese investors invested about $30 billion in early stage US technology research and development through more than 1000 investment agreements. McKinsey's *Artificial Intelligence: The Next Digital Frontier?* report shows that technology giants including Google invested between $20 billion and $30 billion in artificial intelligence in 2016. Ten percent of this was used for artificial intelligence acquisitions, with the remaining 90 percent used for R&D and deployment.

Google's CEO Sundar Pichai believes that Google's business development strategy is shifting from "mobile first" to "artificial intelligence first." Google is the most active buyer in the AI market. Since its first acquisition in 2006, Google's acquisitions in the field of artificial intelligence have reached 18 in recent years. This exceeds the sum of the acquisitions made by Microsoft and Facebook. Among Google's acquisitions are DeepMind Technologies, which developed the Go program AlphaGo, DNN research, which specializes in deep learning and neural networks, and Emu, a

officially launched in 2014 as a new program of research and innovation funding to replace the Seventh Framework Programme and support the Europe 2020 strategy.

smartphone messaging application company. After acquisitions, Google is able to consolidate and integrate the acquired technology into the company's products. For example, Emu is used in the Google Hangouts and Google Now products, and DNN research has greatly improved Google's image search function. At the same time, Google has increased its R&D investment and established artificial intelligence–related funds, research institutes, and workplaces to support artificial intelligence research and development around the world.

Apple has also been increasing its acquisitions in recent years. So far, it has made eight acquisitions, ranking second among all the giants. In 2016 alone, Apple acquired three artificial intelligence companies: Emotient, Turi, and Tuplejump. In May 2017, Apple acquired Lattice Data, which mainly uses artificial intelligence to process unstructured "dark data," for $200 million. Apple recently launched a website called "Apple Machine Learning Journal." Through the blogs of Apple software engineers, the site records and shares some of their new research and innovations in the field of artificial intelligence and machine learning and displays the company's top artificial intelligence research projects.

Baidu was the first leading enterprise in China to deploy artificial intelligence, entering this market in 2013. During the three years from 2014 to 2016, Baidu spent more than RMB20 billion ($2.8 billion) in research and development expenses. It has established artificial intelligence as the core of its core business. In March 2017, it took the lead in building the "National Engineering Laboratory for Deep Learning Technology and Applications."

The Chinese Government Has Begun to Increase Investment in the Artificial Intelligence Field

Amid the wider trend of artificial intelligence development, the Chinese government has accelerated the introduction of policy documents encouraging and supporting the development of artificial intelligence, urging all functional departments and local governments at all levels to increase policy support for and investment in artificial intelligence. On March 5, 2017, Premier Li Keqiang stated in his 2017 government work report that the government would implement *Made in China 2025* in full; speed up the application of big data, cloud computing, and Internet of Things; and make the development of smart manufacturing a major strategic priority.

He said the government would fully implement development plans for strategic new industries, to accelerate R&D and transformation in industries such as "new materials, artificial intelligence, integrated circuits, biopharmaceuticals, and 5G." This was the first appearance of "artificial intelligence" in the national government work report. In recent years, the national policy has attached great importance to the development of artificial intelligence. In December 2015, the Ministry of Industry and Information Technology issued the *Guiding Opinions on Actively Promoting "Internet Plus" Action Plan*. In February 2016, the Ministry of Science and Technology said at a press conference that an "Artificial Intelligence 2.0" pilot might soon be added to the "Science and Technology Innovation 2030 Megaprojects." In the future, governments at all levels will increase their support for artificial intelligence and accelerate the launching and development of artificial intelligence projects. The artificial intelligence policies are expected to continue to pay dividends, and in 2017, China's artificial intelligence will usher in a truly new era.

In May 2015, the State Council issued the *Made in China 2025* plan. In April 2016, the *Robotics Industry Development Plan (2016–2020)* was issued. In December 2016, the *Three Departments' Notice on Promoting the Healthy Development of the Robot Industry* and standards for the industrial robot industry were released. The introduction of industrial policies such as these has laid a solid foundation for the rapid development of China's robotics industry. In the *Made in China 2025* plan, the government has proposed robots as a key development area. The *Robotics Industry Development Plan (2016–2020)* proposes to form a relatively complete robotics industry system, to secure the continuous growth of the industry, markedly improve the level of the technology, make major breakthroughs in parts and components, and achieve significant results in integrated applications. The *Three Departments' Notice on Promoting the Healthy Development of the Robot Industry* further clarified the requirements: driving the rational development of the robot industry, strengthening abilities in technological innovation, accelerating results and transformation, expanding the industrial robot market, and launching pilots of service robots. At present, over 20 provinces and cities are cultivating robotics as a key industry, and there are over 40 robotics parks built or under construction. The robotics field carries a risk of turning a high-end industry into a low-end one, with overcapacity for low-end products. The setting of industry standards will raise the bar for market entry.

While the Chinese government is investing more, the current US government is reducing the fund for artificial intelligence projects. The Trump administration issued a budget reducing funding for several government agencies that have always supported artificial intelligence research.

China's Artificial Intelligence Companies Strive for the Upper Reaches and Increase Capital Investment

Chinese enterprises have shown no weakness in the middle of the big trend toward artificial intelligence, not only accelerating research and development in technology and applications but also investing significant capital reserves. Computer vision, robotics, and natural language processing are highly regarded subsectors in the Chinese capital market. During 2015–2016, investment in each of these three areas exceeded RMB1 billion ($140 million). Even in the areas of smart homes, smart security, smart driving, and smart finance, investment in each was more than RMB500 million ($70 million).

According to the PricewaterhouseCoopers report, the United States will be the first to enjoy the fruits of artificial intelligence development. In the early stage of artificial intelligence development, North America's productivity growth will be higher than that of China due to relatively high technology maturity and a large number of jobs that can be replaced by this technology. However, after ten years of relatively slow accumulation of technology and expertise, China will begin to catch up with, and overtake the United States.

Artificial intelligence is also challenging traditional industry giants. If companies cannot adapt to the transformations brought by artificial intelligence, they will give up a significant portion of their market share to emerging companies.

Bibliography

http://media.people.com.cn/n1/2017/0802/c40606-29443138.html. [2017-08-14].

https://baijiahao.baidu.com/s?id=1571803370794508&wfr=spider&for=pc. [2017-08-10].

“百度豪赌200亿，IBM每年投入超30亿美元:押注人工智能的先锋如何避免成为先烈?”. 8 August 2017. 11 August 2017. http://www.sohu.com/a/163166220_323203.
“海外巨头"抢食"人工智能 收购频繁加大研发投入”. 2 August 2017. 14 August 2017. http://media.people.com.cn/n1/2017/0802/c40606-29443138.html.
“医疗AI最受资本欢迎， 谷歌是AI最大买家【CB insights百页AI报告】”. 9 August 2017. 10 August 2017. http://www.sohu.com/a/163433981_198516.
“资金人才, 人工智能领先的关键”. 25 May 2017. 10 August 2017. http://china.huanqiu.com/hot/2017-05/10737269.html.

CHAPTER 14

Tangible Hands

When geeks are engaged in the research and development of artificial intelligence "black technology," they often hope to improve the efficiency and functionality of artificial intelligence technology under optimal conditions, but they neglect the difficulties that may arise in complex scenarios or when the technology is abused by people.[1] This has laid the foundation for the potential improper use of artificial intelligence. On the one hand, artificial intelligence systems will be affected by development limitations and data accuracy. It is possible that they will make erroneous decisions in specific situations, resulting in property damage or personal injury, such as traffic accidents involving self-driving cars. On the other hand, the application of artificial intelligence technology may also bring about fairness and discrimination issues. According to a 2016 report by Harvard University's Kennedy School, skin color is highly predictive of the outputs of current artificial intelligence systems for crime propensity prediction, regardless of how the algorithms are adjusted.

If the above problems are not adequately addressed by law and monitoring, they are likely to have an extremely negative impact on human society when the artificial intelligence "singularity" comes. Therefore, an increasing number of countries, regions, international organizations, and industry associations have proposed that supervision and regulation should be extended to artificial intelligence, especially the algorithms and data

[1] The term "black technology" refers to futuristic technology.

Tencent Research Institute et al. (eds.), *Artificial Intelligence*,
https://doi.org/10.1007/978-981-15-6548-9_14

involved in machine decision-making. Measures include strengthening security controls, building artificial intelligence standards and rules, and encouraging public participation in artificial intelligence governance.

The regulation of artificial intelligence technology sometimes requires collaboration and coordination between multiple departments. There will be different modes of regulation—perhaps an independent overseeing body coordinating artificial intelligence law and ethics, and perhaps also a more decentralized model in which competent departments of various industries and government departments at different levels take the lead on specific problems according to their different responsibilities. A coordinating overseeing body can help to understand accurately the direction of industrial development and use artificial intelligence to improve national wellbeing. However, different levels of supervisory bodies are equally important for the division of responsibility in specific fields because they have more professional expertise.

Establish a Coordinating Body for Overseeing Artificial Intelligence

The United States has established a body for overseeing artificial intelligence technology led by the White House Office of Science and Technology Policy (OSTP).[2] In May 2016, the White House established the Machine Learning and Artificial Intelligence Sub-Committee (MLAI), which is responsible for coordinating the research and development of artificial intelligence across departments and proposing technical and policy recommendations on artificial intelligence issues, while monitoring artificial intelligence research and development across industry, research institutes, and governments.

Of the three strategies for artificial intelligence released in the United States in 2016, "Preparing for the Future of Artificial Intelligence" was written by MLAI, the "National Artificial Intelligence R&D Strategic Plan" was written by the Network and Information Technology R&D Sub-Committee (NITRD), and "Artificial Intelligence, Automation, and the Economy," which analyzed economic and employment effects in more detail, was produced by the Executive Office of the President (EOP) and

[2] OSTP is the coordinating body for major US science and technology policies, strategies, and plans, and sits under the Executive Office of the President. It is the only branch of the US government that has technology management as its major focus.

MLAI. In terms of organizational structure, MLAI and NITRD are part of the Committee on Technology (CoT), which sits under the National Science and Technology Council (NSTC). The NSTC is directly subordinate to OSTP and is chaired by the president. The members of the NSTC are the vice president, cabinet secretaries, and other key White House officials.

The EU will also set up a government agency to oversee artificial intelligence. In January 2015, the European Parliament's Committee on Legal Affairs (JURI) decided to set up a working group to study legal issues related to robotics and artificial intelligence. In May 2016, JURI issued its *Draft Report with Recommendations to the Commission on Civil Law Rules on Robotics*. In October of the same year, it released *European Civil Law Rules in Robotics*. Considering that the development of artificial intelligence may bring many new problems, JURI called for the establishment of an EU agency specializing in robotics and artificial intelligence. The agency would be responsible for providing technical, ethical, and regulatory expertise to enable the EU to respond better to the new opportunities and challenges brought by artificial intelligence.

Identify the Role of Different Levels of Regulatory Agencies

Autonomous driving is one of the earliest areas where artificial intelligence is being deployed, and it touches on multiple legal issues such as damages, data leaks, and ethical design. Therefore, we will use this field as an example to analyze the regulatory functions of government agencies.

The United States has always attached great importance to the development of autonomous driving technology. In September 2016, the US Department of Transportation (DOT) issued the *Federal Automated Vehicles Policy*, which clarified the different regulatory powers of the federal and state government departments for the first time. By the end of 2016, 16 states had issued regulations or administrative orders relating to automated driving. These focused on testing licenses and processes, generally supervised by the state's transportation department. The Federal Automated Vehicles Policy further clarified a number of important concepts, including state-level lead regulatory agencies, test conditions, operator requirements, vehicle registration, and insurance requirements. States were encouraged to allow DOT to independently regulate the performance of highly automated vehicles and related technology. At the federal level, the US National Highway Traffic Safety Administration (NHTSA) has extensive enforcement authority over automation technology and equipment. According to congressional directives, NHTSA is obliged to protect public transportation safety, avoid unreasonable risks from motor vehicles or their equipment, and support legislation on autonomous driving at the state level.

Strengthen Safety Controls

Artificial intelligence systems and products can be used by the public only if they are safe. This kind of safety is reflected not only in the quality of the products, but also in the legal and ethical aspects of their production. Many countries have focused on this issue in artificial intelligence–related policies and reports, and hope to ensure safety through various measures.

Pass Safety Tests

As products, artificial intelligence systems and applications inevitably have to take into account product quality; inspection and approval are still necessary steps. Due to improvements in machine learning ability, adaptability, and performance, existing traditional methods cannot be uniformly applied to the inspection and approval of ever-evolving artificial intelligence systems. In the future, testing schemes need to be further systematized to ensure artificial intelligence does not display unnecessary behaviors and operates in line with intended functionality (Scherer, Matthew 2016).

Published on October 13, 2016, the US *National Strategic Plan for R&D of Artificial Intelligence* included a strategy of developing shared public datasets and environments for artificial intelligence testing and training.

Looking ahead, regulators need to establish rules to manage research and testing of artificial intelligence. Such rules should allow artificial intelligence developers to test their designs in a secure environment while collecting data so that regulators can make more informed decisions.

Fair and Transparent Decision-Making

Regulators can only guarantee that law enforcement is justified by ensuring that AI decisions are explicitly and clearly monitored. Policies issued by many countries all mention the importance of transparency of AI decision-making, which can eliminate public distrust of AI technology. In September 2016, the UK House of Commons Science and Technology Committee published a report entitled *Robotics and Artificial Intelligence*, which emphasized the importance of transparent decision-making systems for AI safety and control. According to the report, in key decisions relating to human safety, the lack of transparency in decision-making has become more challenging. For example, there is currently no way for humans to easily follow the decision-making process of an autonomous vehicle.

Algorithmic transparency can increase public trust in AI and should allow humans to test the logic of artificial intelligence. The World Commission on the Ethics of Scientific Knowledge and Technology (COMEST) also released a draft report on robot ethics in 2016, arguing that traceability should be established to ensure that robot decision-making and actions are monitored and controlled.

Minimize "Machine Bias"

Since artificial intelligence is strongly influenced by its human-designed structure and the world view of the data used in the learning process, it is not entirely "fair." The problem of algorithmic bias brought about by artificial intelligence has received widespread attention, and should be reflected in regulatory policies in future. For example, Google Photos marked black people as gorillas, showing how technological errors can lead to harm and possibly social unrest and hatred. According to John Naughton, professor of the Centre for Research in the Arts, Social Sciences, and Humanities at the University of Cambridge, these potential biases have still not been discovered by engineers who firmly believe in the neutrality of science and technology and concentrate on technical functionality rather than the accumulated learning material (Scherer 2016). The above-mentioned report of the UK House of Commons Science and Technology Committee also pointed out that programmers should fully understand their impact on ethics and social sensitivities when writing code that guides everyone's lives.

"Who Encroached on My Privacy?"

When data is integrated into our daily lives and collected, transmitted, stored, and used without our knowledge, we will no longer have any privacy to speak of. This issue will become more apparent in the era of artificial intelligence. The core components of machine learning depend on the daily operation of smart machines, and the data that humans share with artificial intelligence systems becomes no longer private. For example, autonomous vehicles can continuously collect our daily travel patterns, and smart home systems can record our lifestyle and entertainment information at any time. The collection of these data may reveal a more real picture of ourselves, one that does not conform to our expectations.

Google DeepMind and the UK's National Health Service have cooperated extensively in the field of smart medical care, raising concerns among the British people about the way artificial intelligence systems enter, store, and use confidential patient data. Effective measures are needed to tackle these very real challenges and ensure that data used by artificial intelligence systems is reasonably restricted, managed, and controlled to protect privacy rights. In response to such issues, the UK government is working with the Alan Turing Institute to establish a council of data ethics.[3]

Who Is Responsible?

The UN COMEST report explored the complex issue of liability in robotics, that is, "Who should be responsible for the malfunctioning of a robot manufactured through the cooperation of different experts and departments?" This problem is particularly important in the context of the continuous advancement of science and technology, growing market demand, and the increasing freedom and autonomy of robots. According to the report, consideration of robot ethics should not be limited to bodily injury caused by accident or malfunction. It should also include psychological damage caused by intelligent robots, such as robots invading people's privacy and humans being over-reliant on robots because of their human-like behavior. The report proposed two solutions. One was to share responsibility between all those involved in the invention, authorization, and distribution of the robot. The other was to let the intelligent robot take responsibility because of its unprecedented autonomy and ability to make decisions independently.

In fact, perhaps neither of these responsibility attribution methods is perfect, because they ignore the inherent biases of human beings in the process of technological development, as well as the possibility that the

[3] Since this book was published the UK has established the Centre for Data Ethics and Innovation.

technology may be used by people with malicious intentions. To find possible legal solutions, COMEST's report cited Peter Asaro's conclusion that the damage caused by robots and robotics is largely regulated by civil laws related to product liability, because robots are generally considered to be technology products. From this perspective, a large part of the damage caused by robots is attributed to the "negligence" of robot manufacturers and retailers, "the lack of product warnings," and "not fulfilling reasonable duty of care." However, the provisions on product liability in civil law may be difficult to apply to future development, especially in the era of strong artificial intelligence.

Perfecting the Insurance System

Discussions on artificial intelligence liability systems are currently focused on autonomous vehicles and who should be responsible for faults and incidents involving them. In the case of an autonomous vehicle independently making a decision that leads to harm, we are unable to accurately distribute legal responsibility between driver, automaker, and the company that designed the artificial intelligence system. The difficulty of liability determination may lead to a delay in compensation. Therefore, in the future, it is necessary to further improve the commercial insurance system and reduce the risks faced by businesses conducting research and development.

In February 2017, the United Kingdom introduced the new Vehicle Technology and Aviation Bill, aiming to help insurers and insurance companies streamline insurance processes before automated vehicles become more widely deployed.[4] This new insurance rule first solves the problem of compensation for victims. After an accident involving an automated vehicle, the victim will receive compensation from the insurance company in the first instance, and then the problem of attribution of responsibility will be solved. However, this does not bypass the complex liability determination problem, and the driver is still not necessarily excused from liability.

[4] This became an Act in 2018.

The Importance of Anticipatory Regulation

In order to monitor and properly handle the various ethical and legal issues brought about by the advancement of artificial intelligence, such as the design and application of deep learning algorithms and autonomous vehicles, the government should establish a continuous regulatory system. Companies and research institutions are constantly asking government to provide regulatory guidelines and standards, especially with respect to innovative technologies being widely disseminated. Such guidance can allow them to adjust their actions and the future direction of their theory and practice.

One-Size-Fits-All Regulation Is Not Advisable

The government's regulatory approach should be carefully constructed to prevent a one-size-fits-all regulatory approach that hinders technological innovation and application. The United Kingdom's 2016 report on robotics and artificial intelligence mentioned that more transparent regulation can enable the public and manufacturers to make more informed decisions and jointly influence the future development of automated vehicles. Only in this way can the UK market become more suitable for the development of emerging technologies and maintain the UK's position as a leading global economic and technological player. Mike Wilson, representative of the robotics company ABB, expressed concern that the government's regulatory framework was unable to keep up with technological advances. In his opinion, the lack of clear and strict government regulatory rules has created an increasingly obvious regulatory gap, which may deepen the public confidence crisis and hinder the development and application of key innovative technologies in different industries (Gal 2017).

Without Standards, Nothing Can Be Done

Product standards represent a quality commitment to consumers and society on the part of manufacturers, which is essential for protecting consumers' legitimate rights and personal safety. Although standardization systems for most artificial intelligence products have not yet been established, the United States has gained some experience in the construction of federal- and state-level standards systems for autonomous driving.

Similar to the regulatory system for autonomous driving, the United States has adopted an approach of federal coordination and state-level cooperation in the development of standards for automated vehicles. NHTSA is responsible for setting Federal Motor Vehicle Safety Standards (FMVSS) for new vehicles and their equipment, and issuing guidelines to manufacturers. The National Traffic and Motor Vehicle Safety Act clearly says that states may not implement standards that are inconsistent with the FMVSS or performance requirements that are inconsistent with any set by NHTSA. The state traffic management department, as the entity that receives applications for autonomous driving tests, also checks through the application process that the vehicles meet federal standards and performance guidelines.

The Importance of Public Governance

There are active efforts to develop artificial intelligence ethics guidelines in the spheres of industry, academia, and the public, as well as at the national and international level. However, the level of information exchange and participation across the various levels is still unsatisfactory. Similarly, who should oversee the ethical and legal implications of robotic technology and automated systems is still unclear.

The participation of various types of actors in the governance of artificial intelligence issues is of great benefit to broadening modes of thinking and defining the regulatory framework. The US government is already thinking deeply about the shocks and changes that the widespread application of artificial intelligence may bring to society, and organizes various discussions on key related issues. From May to July 2016, the White House Office of Science and Technology Policy co-hosted a series of artificial intelligence workshops with a number of leading universities and research institutions, such as Carnegie Mellon University and New York University's Information Law Institute.[5] Topics covered included artificial intelligence assessment, future development, social and economic impacts, safety and control, and law and governance. The discussions not only established the legal basis and boundaries for the government to manage

[5] A full list of the workshops can be found in: National Science and Technology Council Committee on Technology, Preparing for the Future of Artificial Intelligence. October 2016. https://obamawhitehouse.archives.gov/sites/default/files/whitehouse_files/microsites/ostp/NSTC/preparing_for_the_future_of_ai.pdf [accessed 06/24/2019].

artificial intelligence comprehensively, but also considered in depth the possible deficiencies of artificial intelligence, appropriate safety and control principles, and the long-term impacts on employment, social welfare, and economic development. A series of associated reports were also produced by various institutions.[6]

The artificial intelligence strategy introduced by the United Kingdom also emphasizes the importance of encouraging public participation. The House of Commons Science and Technology Committee's 2016 report on robotics and artificial intelligence pointed out that public participation is conducive to the development of technologies based on artificial intelligence. Only by encouraging the public to interact with and better understand artificial intelligence technology can people become more confident in its future. Public participation can also help regulators better understand and deal with the social problems brought about by artificial intelligence. Many experts believe that more openness in science and technology policy formulation would help promote public participation, while improving understanding of the social, moral, and legal issues brought about by technological development.

China Should Incorporate Artificial Intelligence Regulations Into Strategic Considerations

In recent years, the Chinese government has been vigorously promoting the development of the artificial intelligence industry and strengthening top-level planning, but the focus is still on encouraging technological research and development. There is still a lack of awareness and proactive action with regard to the economic and social shocks brought by artificial intelligence technology, and the adjustments that should be made in monitoring and regulation. For example, in the first road test of Baidu's self-driving car in 2015, because the traffic control department did not have any guidance on autonomous driving tests, Baidu could only arrange for the driver to sit at the steering wheel. The traffic control department did not interfere, tacitly approving the car as a human-driven one. Most other autonomous vehicle companies looking to carry out road rests also exist in this regulatory "grey area."

[6] Such as "Artificial Intelligence and Life in 2030." One Hundred Year Study on Artificial Intelligence: Report of the 2015–2016 Study Panel, Stanford University, September 2016. http://ai100.stanford.edu/2016-report.

On May 23, 2016, the National Development and Reform Commission, the Ministry of Science and Technology, the Ministry of Industry and Information Technology, and the Central Cyberspace Affairs Commission jointly issued the *"Internet Plus" and AI Three-year Implementation Plan*, focusing on building platforms, nurturing enterprises, establishing markets, stimulating innovation, and so on. Although the document also talked about financial support, the construction of a standards system, intellectual property protection, talent training, and international cooperation, there was basically no mention of regulatory principles, the regulatory system, and the construction of relevant regulatory agencies. Therefore, in order to promote the development of China's artificial intelligence industry, relevant legal, ethical, and supervisory systems should be established in parallel. Otherwise, China may end up with a mature industry unable to actually apply its technologies.

Bibliography

Gal, Danit. 英国人工智能的未来监管措施与目标概述 (人工智能各国战略解读系列之六). 孙那, 李金磊, 译.电信网科技, 2017(2). (*Overview of the UK's artificial intelligence future regulation measures and objectives (Analysis of national artificial intelligence strategies, Number 6)*. Translated into Mandarin by Sun Na and Li Jinlei, Telecommunication Network Technologies).

Scherer, Matthew U. "Regulating Artificial Intelligence Systems: Risks, Challenges, Competencies, and Strategies". *Harvard Journal of Law & Technology*, 29: 2 (2016).

CHAPTER 15

Kind AI

In his 1942 short novel, science fiction writer Isaac Asimov proposed three laws of robotics—engineering safety measures and built-in ethical standards to ensure that robots will be kind to humanity and to help people avoid a machine doomsday. First, a robot cannot injure a human being or through inaction allow a human to be harmed; second, a robot must obey orders, unless these orders violate the first law; and third, a robot must protect itself, as long as this does not violate the first or second law.

However, in May 2017, in a Brookings Institution driverless car seminar, experts discussed what a driverless car should do in times of emergency. If the car suddenly brakes to protect its passengers, what about the vehicle following closely behind it? Or when a vehicle swerves suddenly to avoid a child, what about the other people nearby that it hits? With the constant development of AI technology, similar ethical dilemmas will very soon affect the development of various types of artificial intelligence. There are some problems that we are already faced with now. Countries' strategies are mentioning the many ethical issues that artificial intelligence has already given rise to or may give rise to. In addition, businesses and organizations engaged in artificial intelligence technology and research have adopted a variety of countermeasures.

Tencent Research Institute et al. (eds.), *Artificial Intelligence*,
https://doi.org/10.1007/978-981-15-6548-9_15

Ethical Issues Become Artificial Intelligence's Most Formidable Challenge

Currently, the ethical issues relating to AI mainly appear in the following four areas:

Algorithmic Discrimination

Algorithms themselves are relatively objective mathematical expressions, not as prone as human beings to prejudice, emotions, and external factors. However, in recent years, algorithms have also given rise to problems of discrimination. For example, a criminal risk assessment algorithm called COMPAS, used by some US courts, has been proven to systematically discriminate against black people. A black man who has committed a crime is more likely to be incorrectly marked by the system as having a high risk of offending, and thus being sentenced to imprisonment or a longer prison term by the judge, even if he should have been given a suspended sentence. Other examples include some image recognition software identifying black people as "gorillas," and Microsoft's Twitter chatbot Tay, which started making sexist and racist comments in the process of interacting with users in March 2016. As algorithmic decision-making becomes more and more common, similar cases of discrimination will also become more and more common.

Some recommendation algorithms may be innocuous, but if algorithms are used for applications such as criminal risk assessment, loans, and recruitment, they may affect the interests of a whole group or race of people. A small mistake or discrimination in algorithmic decision-making can be exacerbated in the subsequent decision-making, possibly causing a chain reaction of errors. In addition, deep learning is a typical "black box"; even the designer may not know how the algorithm makes decisions, so discovering discrimination and its causes in a system may be technically difficult.

Privacy

Many AI systems, including deep learning, require a lot of data to train the learning algorithm. Data has already become the new oil of the AI era, but this has brought new privacy concerns. One aspect is that the large-scale collection and use of data for AI, especially sensitive data, can threaten privacy. For example, medical and health data may be leaked, impacting individual privacy. How to protect personal privacy in the deep learning process is a very important issue right now.

The wide application of user portraits and automated decision-making may also have an adverse impact on the rights of the individual. Moreover, given the large amount of data transactions between various services, data flows more and more frequently, which could weaken individuals' control and management of their personal data. Of course, there are already some tools that can be utilized to enhance the protection of privacy in the era of AI, such as privacy by design, default privacy, personal data management tools, anonymization, pseudonymization, encryption, and differential privacy. These standards are all constantly developing and improving, and are worth promoting in deep learning and AI product design.

Responsibility and Safety

Some well-known figures such as Stephen Hawking have stressed the need to be alert to strong artificial intelligence or superintelligence that could threaten the survival of humanity. But AI safety usually refers to the safety and controllability of intelligent robots, including behavioral safety and human control. From Asimov's three laws of robotics to the 23 principles of artificial intelligence put forward at the 2017 Asilomar conference, AI safety has always been a focus of attention. In addition, the topic of safety is often accompanied by responsibility. As driverless cars will also have accidents, and intelligent robots will cause damage to people and property, who will bear responsibility? If we follow existing legal liability rules, developers cannot predict everything that may happen in a highly autonomous system, and with the presence of the black box, it can be very difficult to explain the cause of an accident. Thus the future may bring a liability gap.

Robot Rights

How to define the humane treatment of AI? As intelligent autonomous robots become more and more powerful, what kind of role should they play in human society? Is it possible for them to receive the same treatment as humans in some respects, that is, to enjoy certain rights? Can we abuse, torture, or kill a robot? What is the legal status of an autonomous intelligent robot? Natural person? Legal person? Animal? Thing? The EU has been considering whether or not to confer intelligent robots with the legal status of "electronic person," with rights, obligations, and responsibility for their actions.

Government and Organizational Strategies

In addition to seeing the beneficial impacts brought by artificial intelligence, many countries and organizations are beginning to face the ethical challenges posed by the technology and consider possible responses. The AI strategies of major countries all consider issues of AI ethics, with many organizations also beginning to study and address these difficult problems.

United Nations AI Policy

A 2017 report by UNESCO's World Commission on the Ethics of Scientific Knowledge and Technology (COMEST) focused on the ethical challenges posed by contemporary robotics.

Although artificial intelligence robots are generally considered to be the carriers of artificial intelligence systems, their movement functions and applications, as well as machine learning capabilities, make them automated, intelligent electronic entities. Automated, intelligent robots are not only capable of complex decision-making processes, but can also execute real-world activities through complex algorithms. The evolution of these new capabilities is in turn leading to the emergence of new ethical and legal issues. Specifically, there are four main aspects:

Challenges Brought by the Use of Robots

The COMEST report cites a report on artificial intelligence and robotics issued by the European Parliament's Committee on Legal Affairs in 2017, which expressed concern about the risks robots will pose for humanity including safety, privacy, integrity, dignity, and autonomy. To address these challenges, measures and design principles put forward by the committee included security and privacy by design; use of the precautionary principle; testing robots in real-life scenarios; obtaining informed consent before man-machine interaction; opt-out mechanisms (kill switches); and starting an inclusive debate on the sustainability of current tax and social systems.

Robotics and Roboethics

Roboethics deals with ethical issues involved in the human construction and deployment of robots, as distinct from machine ethics, which deals with how to program robots with ethical codes. The COMEST report recognized that there are no universally accepted ethical codes specifically for roboticists, but gave an overview of some approaches used in different countries. For example, the South Korean government established a working group to draft a roboethics charter, while Japan started work on a set of guidelines for the deployment of robots, which included the use of a central database to which robots would report incidents involving harm to humans.

Toward a New Mechanism for the Allocation of Responsibility?

The COMEST report explored the complex problem of who would bear responsibility for harm caused by a robot, given that many parties may be involved in the robot's design, construction, and use. The report suggests that there seems to be shared responsibility, but that at the same time this tends to dilute the notion of responsibility altogether. Another solution is to make intelligent robots take responsibility, because they do have unprecedented autonomy and the ability to make independent decisions.

The Importance of Traceability

The COMEST report argues that in the ethical and legal regulation of robots and robotic technology, traceability is a crucial issue. Traceability is important to allow human regulatory bodies to understand the thinking and decision-making process of intelligent robots. This is needed for conducting comprehensive post-incident investigations, enabling litigation, and making necessary corrections.

US Robotics Roadmap

On October 31, 2016, over 150 experts in the United States collectively completed "A Roadmap for US Robotics—From Internet to Robotics."

Although the roadmap is a technology-related document, the authors were aware that the development of robotics in the United States and elsewhere involves non-technical challenges in areas such as law, policy, ethics, and the economy. The roadmap presented the more pressing of these challenges and listed some examples of efforts to address them. It did not aim to cover all the issues or to express a consensus on what robotics policy should be, but rather to raise some important challenges recurring in the literature. In addition, it committed to participate in and support similar dialogue on these issues, which would by necessity be interdisciplinary. The main issues addressed included safety, liability, impact on labor, social interaction, personal privacy, and data security. After discussing these issues, the report made the following recommendations:

First, all levels of government should continue to accrue expertise in cyber-physical systems, to foster innovation in robotics, maximize its potential for social good, and minimize its potential for harm.

Second, government and academia should actively cooperate and break down silos between expertise, as few issues can be resolved through reference to only one discipline.

Third, to eliminate research barriers, independent researchers should be assured that efforts to understand and verify systems for accountability and safety purposes do not carry legal risk.

The EU's Robotics R&D Program

In 2014, the EU launched SPARC, an R&D program for robotics. The program adopted a public-private partnership model, driven by the cooperation of the European Union and the European Robotics Association. It includes analysis of the impact of robotics development on ethical, legal, and social (ELS) issues.

The program believes commercial interests, the interests of consumers, and technological advances will lead to the widespread proliferation of robotics in our daily lives, from manufacturing to civil security, from autonomous transport to robot companions. Establishing an early understanding of ethical, legal, and social (ELS) issues will help with undertaking timely legislative action and social engagement. Ensuring that designers of robot systems understand the importance of equality, and providing guidance in the creation of systems that comply with law and ethics, will be key to solving these important issues, and will help to build confidence and support the development of new markets. ELS issues will significantly affect whether robots and robotic equipment become part of our daily lives. To some extent, they will have an even greater impact than technology readiness level on the delivery of robotics systems in the market.

When it comes to ELS issues, SPARC notes that we should take into account not only existing national and international law, but also different ethical and cultural viewpoints, and the rights and social expectations in different European countries. To make the robotics industry aware of these issues, it is necessary to strengthen interdisciplinary education and the construction of legal and ethical infrastructure as the industry develops. It is increasingly recognized that the guaranteeing and enacting of standards, norms, and legislation will become part of the design process for robotics equipment and technology systems.

Organizational Responses

Aside from national government strategies, businesses, associations, and other organizations that work on or are concerned with artificial intelligence are also paying attention to the ethical issues facing AI, and adopting various measures to study and respond to these issues.

Cooperating or Establishing Alliances

Several Silicon Valley giants launched Partnership on AI, which is a nonprofit organization founded by Amazon, Google, Facebook, IBM, and Microsoft. The Partnership is committed to resolving issues such as the reliability of AI technology. Apple later joined the Partnership, and the first formal meeting of the Board of Directors was held in San Francisco on February 3, 2017.

There is also an organization called OpenAI, which is committed to the development and promotion of open source artificial intelligence systems that benefit all of humanity. Google researcher Peter Norvig has emphasized that machine learning research should be disseminated through publications and open source code for the benefit of everyone.

Set Up Special Bodies for Researching and Overseeing AI Ethics

In 2014, Google acquired British artificial intelligence startup DeepMind for $650 million. One of the conditions DeepMind founders attached to the deal was the establishment of an ethics board by Google. This seemed to mark the arrival of a new era of responsible development of AI. However, since the establishment of Google's AI ethics board, Google and DeepMind kept the members and work of the board a closely guarded secret. They refused to publicly confirm the members of the board, despite constant questioning from journalists. They have not disclosed any information about how the board operates.

Eric Horvitz, who was founding co-chair of the Partnership on AI and Director of the Microsoft Research Lab at Redmond, said Microsoft and other companies had established ethics committees years ago to guide research and development. In 2016, Microsoft also created its own AI ethics committee, Aether, and linked it with the Partnership on AI. Horvitz hopes that other companies too would follow their example. He added that Microsoft has already shared best practice in establishing ethics committees with a number of peer companies.

Put Technical Restrictions on AI

Georgia Institute of Technology computer scientist and roboethicist Ronald Arkin noted that for robots to help soldiers or kill people, a robot should not be sent to the army and be left to find out which rules it should follow. If a robot must choose whether to save a soldier or chase the enemy, it must know in advance what it should do. With the support of the US Department of Defense, Arkin is designing a program to ensure that military robots can perform tasks in accordance with international treaties. A set of algorithms called an "ethical governor" will assess whether or not a robot should be allowed to perform a given task, such as firing a missile.

Currently, the United States, Japan, South Korea, Britain, and other countries have substantially increased funding for research and development of military robots. A British expert has claimed that within 20 years automatic killing machine technology could be widely used. The United Nations Convention on Certain Conventional Weapons has heard views on killer robots from technical and legal experts. Peter Asaro, co-founder of the International Committee for Robot Arms Control and Affiliate Scholar at Stanford's Internet and Society Center, said that more and more people agree that robots killing people without the supervision of a human is unacceptable.

How to build ethical robots will have a major impact on the future development of robotics. Liverpool University computer scientist Michael Fisher believes that rule-bound systems will make the public feel at ease. "If they are not sure what robots will do, they will be afraid of robots," he said. "But if we can analyze and prove the reasons for their behavior, we can overcome this problem of trust." In a government-funded project, he worked with Winfield and others on verifying the correctness of a machine's ethical decision-making.

Publish Official Research Reports and Guidelines

The British Standards Institute (BSI), Britain's national standards body, released a set of ethical guidelines for robots in 2016. BSI has over 100 years of history and is highly authoritative worldwide. Its "Guide to the ethical design and application of robots and robotic systems" is mainly aimed at robot designers, researchers, and manufacturers. It provides guidance on how to identify and address ethical hazards associated with robots and robotic systems. The ultimate goal is to ensure that intelligent robots made by humans are able to integrate into the existing moral codes of human society.

In 2017, MIT Media Lab and Harvard University's Berkman Klein Center for Internet & Society launched a $27 million Ethics and Governance of Artificial Intelligence Initiative. They want to solve the human and moral problems arising from artificial intelligence, study how it should take social responsibility (such as ensuring fairness in education and justice), and help the public understand the complexity and diversity of artificial intelligence.

The IEEE Global Initiative for Ethical Considerations in Artificial Intelligence and Autonomous Systems is an industry connections program launched by the IEEE Standards Association in April 2016, as part of IEEE's broader program on ethics, TechEthics. The main aim of the initiative is to enable artificial intelligence and autonomous systems to be built in line with the values of users and society. In this way, we can make increasing human well-being the primary goal against which to measure progress in today's "algorithm era."

The initiative has two results: the "Ethically Aligned Design" document and standards proposals that could become actual operational standards adopted by industry and designers. The first edition of "Ethically Aligned Design" was published on December 13, 2016. It represented the collective wisdom of more than 100 leaders in artificial intelligence, robotics, law, ethics, philosophy, and policy, across academia, science, government, and business. The goal was to provide insight and recommendations that will act as a key reference for artificial intelligence and autonomous systems technicians in the next few years.

CHAPTER 16

The Fight for Talent

In 2017, human beings were once again profoundly shocked by artificial intelligence. The "Master," the mysterious flag bearer under Google's AlphaGo, swept away humans with a record of 60 wins, 0 losses, and 1 tie. The comprehensive victory of "Master" indicates that the era of artificial intelligence is approaching rapidly. Success can be achieved only by capable people. Talent has always been the primary resource for economic and social development. Faced with the rapid development of artificial intelligence, competition for talent has become the most important topic in the field of artificial intelligence. The development of artificial intelligence is inseparable from the unearthing and cultivation of talents. Talents determine R&D capabilities. The world needs excellent artificial intelligence talents to further release the enormous potential of computing and machine learning technologies. A computer program capable of defeating world champions many times at Go is undoubtedly a coup for the fast-growing field of artificial intelligence. Under this surging surface, a more desperate gamble is quietly brewing—the fight for talent in the field of artificial intelligence. In the future, as countries establish more sophisticated artificial intelligence and other scientific personnel training systems at the domestic and international levels, fierce fights are expected to cool down, and artificial intelligence can achieve healthier, more creative, and sustainable development.

Tencent Research Institute et al. (eds.), *Artificial Intelligence*,
https://doi.org/10.1007/978-981-15-6548-9_16

The Fight for Talent Has Fully Commenced

At present, all countries in the world place great emphasis on the importance of scientific and technological talents for the progress of artificial intelligence technology, and try to seize the strategic advantage in the development of artificial intelligence through the possession of scientific and technological talents. Whether it is the US government's three artificial intelligence reports, the Japanese government's new robotics strategy, the British government's 2020 robotics and autonomous systems development strategy, or China's "'Internet Plus' and AI Three-Year Implementation Plan," all invariably emphasize the importance of cultivating scientific and technological talent. In addition to the cultivation of scientific and technological talent, countries also plan to introduce policies and measures to help the existing low-skilled labor force transfer to high-skilled industries, strengthen vocational training, and reduce the impact of artificial intelligence on the job market.

United States: Better Grasp the Needs of National Artificial Intelligence R&D Talent

The United States has always paid attention to the research and development of artificial intelligence, and in recent years its efforts in this area have been accelerating. From October to December 2016, the US White House successively issued three artificial intelligence development reports: "Preparing for the Future of Artificial Intelligence", "National Artificial Intelligence Research and Development Strategic Plan", and "Artificial Intelligence, Automation and the Economy". Among them, Strategy 7 of the "National Artificial Intelligence Research and Development Strategic Plan" mentions that the development of artificial intelligence requires a strong artificial intelligence researcher community. It is necessary to understand better the current and future demands of artificial intelligence R&D talent to ensure that there are sufficient artificial intelligence experts to respond to the strategic R&D areas outlined in this plan. Countries with strong capabilities in R&D are also bound to occupy a leading position in future development. Reports from commercial and academic institutions show that there is a growing insufficiency in the field of artificial intelligence professionals. For this reason, high-tech companies are continuously increasing their investment in the use of artificial intelligence, and universities and research institutions are constantly recruiting artificial intelligence professionals. In the future, the plan states that the United States needs more data on the supply of and demand for national artificial intelligence research and development talents, including the needs of scientific research institutions, governments, and industries. This will help predict future labor force requirements and formulate reasonable plans.

The report "Artificial Intelligence, Automation and the Economy" provides an in-depth look at the impact that AI-driven automation will have on the economy and proposes a national-level response strategy. The report mentions that it is necessary to strengthen the training of artificial intelligence talents and fully understand the situation of practitioners. Formulate research methods and programs, increase the collection of official statistics to reflect the current status of artificial intelligence practitioners, and effectively predict future labor force demand and supply. Build a sufficient and active talent team, increase education and training opportunities related to artificial intelligence, and create and maintain a healthy national artificial intelligence research and development talent team. At the same time, add subjects such as ethics, safeguards, privacy, and security to the artificial intelligence courses of various universities. In relation to government, maintain training of government workers and establish intergovernment personnel exchange programs. Through a series of personnel appointments and innovations in exchange models, train government staff to have a full understanding of the current state of artificial intelligence development, and introduce artificial intelligence into the federal employee training pipeline.

Japan: Cultivating Teams of Professional Talent

In January 2015, Japan's Ministry of Economy, Trade and Industry released the "Robot Strategy: Vision, Strategy, Action Plan," presenting its three core strategies for realizing the robot revolution. The first is to become the world's robot innovation base, completely consolidating the cultivation capabilities of the robotics industry. This involves increasing research collaborations across industry and academia, and increasing the opportunities for users and manufacturers to connect and spark innovation. At the same time, promote work such as the cultivation of talent, next-generation technology research and development, and international standardization. The second is to build the world's first robot application society, making robots visible everywhere. The third is to stride towards a global leadership position in the new era of robotics. Human resources are the key resources highlighted by Japan in the "Robot Strategy." They are the labor force guarantee for the comprehensive realization of the robot strategy. The "Robot Strategy" pointed out that talents in information technology, such as those with skills in robot system integration and software, should be nurtured as part of the crucial cultivation of the robot revolution. One approach is based on cultivating system integrators through practical training and projects that increase the opportunities for them to perform actual on-site installation of robots. Second is the use of vocational training and vocational qualification systems to support the cultivation of system integrator talents, the education and training of talents related to research institutions or universities, and support policies for new entrepreneurial talents, and so on. It is also necessary to develop a policy for nurturing and installing robots based on a medium- to long-term perspective.

UK: Full Scholarship Programs to Promote Science and Technology Education

Demand for skilled workers in robotics and autonomous systems and related fields is growing, but the United Kingdom does not have sufficient trained human resources to achieve its ambitious development goals. The government has promoted a full scholarship program for science and technology education at the level of higher education, which has alleviated this problem to some extent. The 2020 robotics and autonomous systems development strategy points out that this is crucial for promoting the development of artificial intelligence and related technologies and the use of foreign funds to develop the domestic market. Despite this, British universities may suffer harm because talent is sucked away by highly profitable industries, and universities may be forced to change their research directions and shift their focus from necessary exploratory research to more commercializable and profitable areas. The development strategy mentions that the government must strengthen investment in vocational training so that workers can gain new and relevant skills, reduce the negative impact of the large-scale application of automation technology and automated machines on the employment of workers, and stabilize the job market. At the same time, establish an adaptive and timely training program to enable workers to keep up with the latest technological development trends and provide them with life-long learning opportunities as they are forced to change jobs. The report on Robotics and artificial intelligence issued by the House of Commons Science and Technology Committee expresses disappointment for the government's lack of leadership in this area, and calls for the publication of a national digital strategy as soon as possible to help workers better cope with the increasingly automated and autonomous market while preventing the phenomenon of digital exclusion.

China: A High Degree of Emphasis on Cultivation of Talent in Artificial Intelligence Domain

In order to implement the *Guiding Opinions on Actively Promoting 'Internet Plus' Action Plan* and accelerate the development of the artificial intelligence industry, the National Development and Reform Commission, the Ministry of Science and Technology, the Ministry of Industry and Information Technology, and the Cyberspace Administration of China jointly issued the "Notice regarding the 'Internet Plus' and AI Three-Year Implementation Plan" (hereinafter referred to as "Proposal"). The "Proposal" clearly clarifies the need to nurture and develop emerging industries for artificial intelligence, promote smart product innovation in key areas, and increase the level of intelligentization of end products. The government will also provide guarantees for funding, standards systems, intellectual property rights, personnel training, international cooperation, and organization and implementation. The "Proposal" states that relevant research institutions, institutions of higher learning, and experts are encouraged to carry out training in basic knowledge and application of artificial intelligence. Relying on major national talent projects, China will speed up the training and introduction of a group of high-end and interdisciplinary talents. Improve AI-related majors and curricula in universities, focusing on the integration of artificial intelligence with other disciplines, encouraging cooperation between universities, research institutes and enterprises, and building a number of artificial intelligence training bases. Support high-end talents in the field of artificial intelligence to go abroad to conduct academic exchanges on issues of cutting-edge technologies and standards.

Whoever Obtains Artificial Intelligence Talents Obtains Everything Under Heaven

The conflagration of the strategic contest over talent has caused smoke to spring up everywhere. On July 6, 2017, LinkedIn, the world's largest social networking platform for the workplace, released the industry's first "Global AI Talent Report". This report is based on LinkedIn's data on 500 million global high-end talents, as well as its series of in-depth analyses of the status, trends, and supply and demand for core technical talents in the global AI field. According to the report, as of the first quarter of 2017, there are more than 1.9 million technical talents in the global AI field based on the LinkedIn platform. Of these, the highest number (over 850,000) are US-related, with China-related talents exceeding 50,000, placing it seventh in the world. In the past three years, the number of AI positions posted through the LinkedIn platform has soared from 50,000 in 2014 to 440,000 in 2016, an increase of nearly eight times. Demand is currently highest for the talents in the layer of AI fundamentals, especially in terms of algorithms, machine learning, GPUs, smart chips, and so on, exhibiting a significant talent gap compared with the technology layer and the application layer.

Global giants are aware that the core of competition in the artificial intelligence field is the competition for talent. In the current environment, major technology giants such as Facebook, Google, Amazon, and Microsoft have used talent discovery as a core strategy for the development of artificial intelligence. While investing heavily in the development of artificial intelligence business, competition for talents in this field has also heated up. From the top-level labs in academia and industry to graduates from colleges and universities around the world, all are long-term battlefields where technology companies seize and store up talent resources. A predatory strategy is the universal choice for these technology companies. With first-class talents and first-rate treatment, such technology companies have become the gathering places for cutting-edge talents in the industry. Talents promote the rapid development of enterprises, and continuous growth in performance has in turn prompted companies to continue to attract a large number of outstanding talents to form a virtuous circle. Last year, technology companies including Google,

Facebook, Microsoft, and Baidu spent about USD 8.5 billion for acquiring and recruiting talent, which is four times higher than in 2010. US companies pay USD 650 million a year for 10,000 talents in artificial intelligence. Among them, Amazon spent more than USD 200 million US dollars to recruit artificial intelligence talent, ranking first among major companies. Facebook's strategy for attracting talent includes providing up to hundreds of thousands of dollars in salary, scattering work locations around the world, and more. In addition, artificial intelligence is not only favored by technology giants. Some startups also choose artificial intelligence as a breakthrough point, which means that the competition of artificial intelligence talents has gradually extended from the technology giants to startup companies. The data shows that in the past one or two years when the overall entrepreneurial situation was not optimistic, over 60% of artificial intelligence companies still obtained the support of venture capital.

With the changes in the international and domestic economic situation and the rapid disappearance of the demographic dividend, the Chinese economy urgently needs to find new growth engines. The huge productivity-enhancing potential brought about by smart applications based on artificial intelligence is broadly viewed as positive by all sectors of society. Domestic technology companies such as Baidu, Alibaba, and Tencent are leaders in the industry. The flow of talents in the three companies has always been extremely frequent, and talents are constantly coming and going in the chain of resources. In order to fully compete for talents, the three companies have successively established their own artificial intelligence talent research institutes, focusing on cultivating and unearthing local talents. In this regard, leading one-stop big data recruitment platform eCheng's Talent Big Data Research Institute, based on the data held by the eCheng platform in April 2017 relating to Baidu, Alibaba, and Tencent, released the "BAT Artificial Intelligence Domain Talent Development Report". The talent strategies of the above three companies are valuable to reference. The report shows that in the total artificial intelligence talent pool, Baidu leads, and in terms of salary and stability for artificial intelligence talents, Alibaba seizes the remuneration high ground, while Tencent has the most stability. Data analysis, data mining, language recognition, and natural language processing jobs are all areas where these three companies must fight for talents. The artificial intelligence talent structures of the three companies are basically centered around these core businesses. In accordance with different core business operations, the functional layout of artificial intelligence talents also has different areas of focus in each company: Baidu stresses search, Alibaba is best at strategy, and Tencent emphasizes analysis. Baidu, as the leader in domestic search engines, has an ample talent pool in terms of algorithms, architecture, and so on; Tencent focuses on products, so the proportion of technical talents is relatively small; and Alibaba's e-commerce background determines that the current proportion of the company's personnel conducting technology R&D is not high. Baidu is currently playing the role of the "Whampoa Military Academy"[1] for domestic talents in artificial intelligence. Alibaba leans towards attracting talents with high pay, and Tencent is steadily achieving an efficient output ratio from its talent. Among the three companies, Baidu is the current leader in the artificial intelligence talent pool. With relatively low salaries and higher job-hopping aspiration, its artificial intelligence talents are welcomed by the market and are more prone to be poached. Alibaba has the most advantage in terms of salary level and the scale of salary increases. Given Alibaba's late start relative to Baidu, adopting a strategy of high pay to obtain high-quality talent is a common method used by many chasers in the field of artificial intelligence in recent years. Tencent's algorithmic strategy, engineering, and data analysis roles have average employment times of more than three years, and salaries remain in the middle of the three BAT companies. Talent retention and budgeting are well controlled, and it is steadily realizing its strategic positioning with respect to artificial intelligence. In terms of the source of artificial intelligence talents, among domestic universities, graduates from 20 universities, including Peking University, Tsinghua University, Beijing University of Posts and Telecommunications, Huazhong University of Science and Technology, and China University of Science and Technology are more popular with BAT. Computer science majors getting employment in artificial intelligence is the common road, and a Master's degree has become the average threshold for entry.

[1] This is a military school that produced many prestigious commanders who fought in many of China's conflicts in the twentieth century.

Bibliography

"BAT人工智能领域人才发展报告". 12 June 2017. 3 July 2017. https://www.ifchange.com/operation/bat

"领英发布《全球AI领域人才报告》，揭示全球AI人才图谱". 10 July 2017. 12 July 2017. http://www.sohu.com/a/155984435_133098.

"人工智能人才争夺战持续升级". 4 May 2017. 3 July 2017. https://www.jiqizhixin.com/articles/2017-05-04-6

PART IV

Law: Fairness and Justice in the Age of AI

The continuous development of artificial intelligence technology has brought shocks and challenges to the existing legal system. When poems created by artificial intelligence resonate with us, when we see autonomous cars driving on the road, when our lives are no longer lonely because of the presence of companion robots, we also need to address how to adjust the existing legal system to regulate and promote the future development of artificial intelligence. The law is accustomed to lagging behind the development of new technologies, but in the field of artificial intelligence, whether we need to create some forward-looking legislative frameworks, and what they should look like, are legal problems that all countries in the world need to tackle together.

CHAPTER 17

How To Be Accountable for AI?

On May 7, 2016, 40-year-old Joshua Brown was driving a Tesla Model S at full speed in autopilot mode on a Florida highway when he hit a white tow truck crossing his path. The issue that everyone is generally concerned about is that since it was on autopilot, who should bear the legal responsibility after the accident? Is it possible for AI or autonomous systems to take on accountability?

The Dilemma of Traditional Liability Theory: Can the Old Bottle Still Be Filled with New Wine?

The attribution of legal liability is the primary legal challenge facing the development of artificial intelligence. This involves how to ensure that artificial intelligence and autonomous systems are held accountable. Establishing legal responsibility requires investigation, the protection of the legal rights of the relevant subjects, and the maintenance of social relations and social order regulated by the law. In the Tesla case, the US National Highway Traffic Safety Administration (NHTSA) ultimately came to the conclusion that Tesla's autopilot design was not significantly flawed. However, NHTSA did not give any definitive conclusions as to how the legal liability for automated driving accidents is defined. NHTSA states that its monitoring of the reliability of automated driving functions is not over and reserves the right to intervene again when necessary.

Tencent Research Institute et al. (eds.), *Artificial Intelligence*,
https://doi.org/10.1007/978-981-15-6548-9_17

From the perspective of traditional liability theory, legal liability is divided into fault liability and no-fault liability. Fault liability will take "fault" as a constitutive requirement of liability, and also as the ultimate condition, so if there's no fault, there is no liability. Fault liability is the most common form of legal liability, and holds a dominant position in legal liability; traditional tort law is mainly based on the principle of fault liability. However, after entering the era of artificial intelligence, artificial intelligence systems can already do some work independently without human operation and supervision, and how to judge and determine liability for the damage caused by machine autonomy is a big challenge. In the case of the Tesla incident, at least from the current survey of the situation, the driver and car manufacturer were both not at fault, but the accident still happened. There needs to be someone to take responsibility, and under such circumstances, how to define the liabilities of all parties leads to dilemmas.

In view of the fact that the problem of attributing liability for incidents involving artificial intelligence has already emerged in practice, especially in the application of autonomous driving and robotics, there is an urgent need to address this problem. Some countries and regions have begun to explore the issue at the legislative level, and the international community has also started actively exploring and debating it.

In December 2016, the Institute of Electrical and Electronics Engineers (IEEE) released "Ethically Aligned Design: A Vision for Prioritizing Human Well-being with Autonomous and Intelligent Systems (AI / AS)." The second basic principle proposed is the principle of responsibility. The document points out that in order to solve the fault problem and avoid public confusion, the AI system must be accountable at the procedural level to prove why it operates in a certain way.

Legislative Attempts in the Field of Autonomous Vehicles

On the morning of July 5, at the 2017 Baidu AI Developers Conference, Baidu founder, chairman and chief executive officer Robin Li streamed a live video of a self-driving car developed by the company. In the video, Robin Li sat in the passenger seat of a red car with no driver in the driver's seat. In response, the Beijing Traffic Management Bureau announced that it is actively carrying out investigations and verifications and that it supports technological innovation in self-driving cars, but such innovation should happen legally, safely, and scientifically. Many people think that Robin Li's behavior was inappropriate, but no regulations have yet been issued in China regarding autonomous vehicles.

Global legislation for autonomous driving, which is the domain of artificial intelligence most widely applied in this current stage, keeps pushing forward. International organizations such as the United Nations and countries including the United States, Germany, and the United Kingdom are actively revising existing laws and regulations or formulating new laws and policies. The deployment of autonomous driving technology has cleared legal obstacles and made positive progress.

On March 23, 2016, the United Nations "Vienna Convention on Road Traffic" regarding management of road traffic was amended. The amendment clearly stipulates that, under circumstances in which transportation is in full compliance with UN regulations governing the use of vehicles or the driver can choose to turn the technology off, then the responsibility of driving a vehicle can be transferred to automated driving technologies. This means that 72 states which signed the convention can allow cars with automated driving technologies to drive autonomously at specific times, clearing barriers to the use of these technologies in transportation and logistics.

The United States and German legislation in the autonomous driving domain mainly focuses on the definition of liability. The US National Highway Traffic Safety Administration released the "Preliminary Statement of Policy Concerning Automated Vehicles" in 2013. Nine states, including Nevada, California, Florida, and Michigan, also passed autonomous vehicle legislation to stipulate the bearing of liabilities in test incidents, which is: if the testing of a vehicle converted to a self-driving vehicle by a third party leads to the loss of property and casualties, the original manufacturer of the vehicle is not responsible unless there is evidence that the vehicle already had flaws before being converted to a self-driving vehicle. For example, if Google begins tests with a Mercedes-Benz vehicle, the safety liability is borne by Google.

The German Road Traffic Act stipulates that strict liability for road traffic accidents is independent of the degree of automation of the vehicle, that is, motor vehicle owners must assume liability. However, according to predictions by German scholars, this liability will gradually shift from the driver to the manufacturer of the automated driving system as the technology evolves. In 2016, the German legislature initiated legislative amendments to the provisions of the German Road Traffic Act, such as "the driver must remain vigilant throughout the vehicle's driving process" and "the driver's hand cannot leave the steering wheel." In May 2017, Germany adopted an act on self-driving cars to clear barriers to road tests for autonomous vehicles, which states: first, the driver must always sit behind the steering wheel in order to take control of the autonomous vehicle when necessary; and second, on-road testing is permitted and drivers are able to not participate in driving behavior (meaning they can surf the Internet, send mail, and so on); third, a "black box" must be installed to record driving activities; fourth, drivers who participate in driving bear liability according to their duty of care and fault, otherwise the manufacturer assumes responsibility.

The UK Centre for Connected and Autonomous Vehicles (CCAV) has issued two reports proposing recommendations on insurance and product liability. Mandatory motor vehicle insurance was extended to autonomous vehicles to protect victims of autonomous vehicle accidents. Victims will be able to claim indemnities directly from motor insurers and insurers will have the right to recover from entities liable under existing law (such as product liability).

Exploration of Legal Liability for Robots

As intelligent robots become more and more widely used, defining their liability has also aroused great concern and attention from all parties. In August 2016, UNESCO and the World Commission on the Ethics of Scientific Knowledge and Technology discussed the responsibility of robots in the "Preliminary Draft Report on Robotics Ethics" and proposed a feasible solution, namely adopting responsibility-sharing solutions, to let all who participate in the process of inventing, licensing, and distribution of robots bear liability.

The EU has also made positive attempts in the legislation of liability for intelligent robots. As early as January 2015, the European Committee on Legal Affairs (JURI) decided to set up a working group to study legal issues related to the development of robots and artificial intelligence. In May 2016, the Committee on Legal Affairs released the Draft Report with recommendations to the Commission on Civil Law Rules on Robotics. In October of the same year, the "EU Civil Law Rules on Robotics" were released. On the basis of these studies and reports, on February 16, 2017, the European Parliament voted to adopt a resolution recommending that the European Commission submit legislation on robots and artificial intelligence[1], including establishing a European agency specializing in robotics and artificial intelligence to reconstruct liability rules for intelligent robots.

In the opinion of the Committee on Legal Affairs, today's robots already have the autonomy and cognitive characteristics to learn from experience and independently make judgments, and can substantively adjust their behavior. The legal liability arising from infringements by robots has thus become a major issue. The stronger the autonomy of a robot, the harder it is to treat it as a simple tool in the hands of other agents (e.g., manufacturer, owner, user), which in turn makes existing liability rules begin to become inadequate and therefore requires new rules.

The new rules focus on how to make a machine bear liability for some or all of its behavior, and as a result the issue of whether robots should have legal status will become more and more urgent. Ultimately, the law needs to respond to the question of what a robot actually is—whether it should be treated as a natural person, a legal person, an animal or thing, or whether the law should create a new type of legal body for it, with its own rights, obligations, responsibilities, and so on.

Under the current legal framework, a robot itself is not liable for damages caused to third parties by its own actions or negligence. Moreover, established liability rules require robotic behavior or negligence to be attributed to specific legal entities, such as manufacturers, owners, users, who can anticipate and avoid the robot's harmful behavior. Further, the rules on product liability and liability for dangerous goods can make these legal entities take strict liability for the robot's behavior. However, if robots make decisions autonomously, the traditional rules of liability will not be enough to solve the problem of liability of robots, because the traditional rules may not be able to identify the responsible parties and make them pay compensation.

In addition, the shortcomings of the existing legal framework are even more apparent with respect to contractual liability, as robots are now able to choose between contract parties, negotiate contractual terms, conclude contracts and decide whether and how to implement the contracts reached, which make conventional contractual rules inapplicable. In the area of non-contractual liability, the existing rules of product liability can only

[1] In the EU, only the European Commission has the power to propose legislative proposals, but the European Union is not obliged to comply, though if it refuses to do so, it must state its reasons.

cover the damage caused by the manufacturing defects of the robot, and the victim must be able to prove that there is actual damage, product defects, and a causal relationship between the defects and the damage. However, the current legal framework cannot completely cover the damage caused by the new generation of robots, who will learn from their own changeable experiences and interact with their environment in a unique and unpredictable way.

Constructing a Reasonably Structured Liability System

The rapid development and application of artificial intelligence does bring many problems to human society. However, we still have reason to believe that the legal system can control the public danger caused by artificial intelligence without hindering innovation. Therefore, it is very important to define a reasonably structured liability system and clearly define the principal responsibilities and obligations of designers, producers, sellers, and users of artificial intelligence projects. The IEEE elaborated on the measures that different subjects should take with respect to artificial intelligence liability in "Ethically Aligned Design: A Vision for Prioritizing Human Well-being with Autonomous and Intelligent Systems (AI / AS)," stating: the legislature should clarify responsibility, fault, liability, accountability and other issues in the development of artificial systems, in order to make it easy for manufacturers and users to be aware of their rights and obligations; artificial intelligence designers and developers, when necessary, should consider the diversity of cultural norms used by groups; stakeholders should work out new rules when AIs and their impacts go beyond the established ones; and manufacturers and users of autonomous systems should create a recording system that records core parameters.

CHAPTER 18

Deep Privacy Concerns

Online video-streaming company Netflix once released hundreds of millions of items of "anonymously processed" data on moving ratings, preserving only each user's rating of the movie and the timestamp of when they rated it. It hoped to find a better movie recommendation algorithm through the form of a contest. However, in 2009, two researchers at the University of Texas, by comparing these anonymous figures with the open IMDB database, were able to successfully match anonymous data to specific users, and Netflix ultimately had to cancel what was originally planned to be an annual competition. Netflix's case shows that with big data analytics people's secrets cannot be hidden anywhere. So-called privacy protections are nothing more than "the emperor's new clothes."

Privacy and Data Protection Are Core AI Issues

Entering the era of artificial intelligence, with the integration of big data technology and intelligent technology, government and business decision-making increasingly rely on large-scale data collection, analysis, and use, in order to make traditional society more transparent. With the overlay of all three—the Internet of Everything, big data, and machine intelligence—people may no longer have any privacy to speak of.

At the same time, businesses have been exaggerating the great convenience that big data and artificial intelligence bring to people's productivity and lives, and users themselves often overlook the privacy and personal data hazards brought on by new applications of these new technologies.

Tencent Research Institute et al. (eds.), *Artificial Intelligence*,
https://doi.org/10.1007/978-981-15-6548-9_18

Currently, smart apps have become an indispensable tool in people's lives. While providing domestic services, these apps collect a great deal of personal data to push accurate marketing information to users. However, one potential danger of precision marketing is "precision fraud." Many fraud cases show that this will bring great damage to personal and property safety.

As Dr. Wu Jun said, data is the cornerstone of human civilization. Big data has a decisive role in the emergence and development of machine intelligence, but since big data analysis can understand details about people's personal lives or various information within an organization, it can trigger public concerns about privacy rights. The British report, "Artificial intelligence: opportunities and implications for the future of decision making," points out that the ability to protect citizens' data and privacy and treat each citizen's data without discrimination, as well as guarantee the integrity of citizens' personal information, in the course of using citizens' data for analytical purposes, is crucial for the government to win public trust and protect its own citizens. Entering the era of artificial intelligence, privacy and data protection are still core issues that require our attention.

Global Legislation on Privacy and Data Protection Is Heating Up

Overall, personal privacy and data protection have been major concerns of the international community for a long time. Since the enactment of the "Sweden Data Act," the first personal data protection act in 1973, a wave of legislation on the protection of personal information has been released in the world. The United States enacted the "Privacy Act" in 1974, which stipulates rules for the protection of personal information in the public domain. The EU adopted the "Directive on the protection of individuals with regard to the processing of personal data and on the free movement of such data (95/46 / EC)" in 1995 (referred to as the 1995 Personal Data Protection Directive), and Member States immediately converted it into domestic legislation. South Korea, Japan, Singapore, and other countries have enacted personal information protection laws and established basic rules for personal information collection, use and cross-border transmission. As of December 2016, more than 110 countries and regions all over the world have formulated specific personal information protection laws.

In recent years, with the rapid development and application of big data, cloud computing, and new artificial intelligence technologies have brought new challenges to existing legal systems for the protection of personal information. Activities related to national legislation and legal reform have become more frequent in all countries.

In 1995, the EU enacted the "Directive on the protection of individuals with regard to the processing of personal data and on the free movement of such data," which is the fundamental legislation for personal information protection in the European Union area. The EU Member States introduced their own personal information protection laws in accordance with the directive. However, ever-changing information technology has made the main principles and application of the directive very uncertain and has led to major differences between EU Member States in their understanding and enforcement of the personal data protection directive. On January 25, 2012, the European Commission promulgated the "Legislative Proposal on the Personal Data Protection Directive 1995" ("General Data Protection Regulation") as a comprehensive revision of the 1995 Personal Data Protection Directive. On December 15, 2015, the tripartite body of the European Parliament, the Council, and the Commission reached an agreement on EU data protection reform during the final phase of the legislative process. On April 14, 2016, the EU Parliament passed the final version of the ordinance. In the newly adopted regulations, the European Union strengthened personal privacy and data protection. Among them, the provision related to automated decision-making including profiling will have a significant impact on the practices of the big data-based Internet industry. That is, the user has the right to refuse companies' use of automated decision-making and profiling, and the data used for profiling cannot include special categories of personal data, including ethnic or racial origin, political opinions, religious or philosophical beliefs, trade union membership, genetic data, biometrics, health status, and sexual orientation.

In the field of electronic communications, the European Commission announced on January 10, 2017, a proposal to develop a more stringent privacy regulation, the "Regulation on Privacy and Electronic Communications (ePD)." This aims to further strengthen electronic communications data protection. The ePD expands the scope of the affected entities, stipulating that privacy protection rules will also apply to emerging electronic communication service providers such as WhatsApp, Facebook Messenger, and Skype, to ensure that emerging communication service providers and traditional communication service providers are able to provide the same level of protection for users' privacy. ePD also expands the scope of content to be protected, to include communication content and metadata (talk time, location, etc.). Metadata includes highly private content. Except for data used for billing and the like, all metadata needs to be anonymized or deleted if user consent has not been given. ePD also includes: provisions for end-users to protect individual privacy by controlling the sending and receiving of electronic communications; various remedies end users can take; and other important content such as punishments and responsibilities that must be assumed in the case of breaching the regulation.

In addition, Japan, South Korea, and other countries also revised their legislation on the protection of personal information. On March 22, 2016, the Korea Communications Commission (KCC) made drastic amendments to the "Act on Promotion of Information and Communication Network Utilization and Information Protection" and other related laws to further improve the rules on personal information entrustment and increase the requirements for the person responsible for the protection of personal information. It also added rules for the deletion and blocking of exposed personal information, and so on. Japan promulgated the "Amended Act on the Protection of Personal Information" on September 9, 2015, and also added certain anonymous information requirements for the current development of the technology industry, the establishment of a personal information protection committee and rules on cross-border transfer of data and other issues. The new law also places new restrictions on sensitive information, including a prohibition on obtaining and providing sensitive information without the consent of the data subject.

Under the existing legislative framework for the protection of personal information, some countries are actively formulating personal information protection rules for new industries such as cloud computing and big data. The French personal information protection agency, the National Commission on Informatics and Liberty, released the "Guide on the Conservation of Cloud Computing Data", advising on the factors that cloud service agreements should include and the security management of cloud computing. The Japanese government has promulgated the "Guide to Safe Use of Cloud Services," which sets out the issues that cloud customers and cloud service providers should pay attention to in the protection of personal information. Japan's Ministry of Internal Affairs and Communications released the "Guideline for Handling Smartphone User Information (Draft)," which sets out measures to protect the privacy of smartphone users.

In recent years, China's personal information protection legislation activities are also constantly advancing, and have achieved some results. In 2012, the National People's Congress (NPC) Standing Committee adopted the "Decision on Strengthening the Protection of Network Information" and established a number of principles for the protection of personal information. In 2013, it amended the "Consumer Protection Law" to compose relevant regulations on the protection of consumer personal information. In 2009 and 2015, it passed Amendments VII and Amendment IX to the Criminal Law, which specifically increased the penalties for both selling or illegally providing citizens' personal information and stealing or illegally obtaining citizens' personal information. In 2016, the NPC Standing Committee passed the "Cyber Security Law," which synthesized China's past experiences with personal information protection legislation. In view of prominent problems experienced in practice, it confirmed a system using some mature approaches from recent years, and established fundamental rules for the protection of personal data being collected and used in the era of big data. These fundamental rules include legal and legitimate (network operators collecting personal information must have a legitimate purpose and be using appropriate and legal methods); informed consent (network operators are required to disclose privacy rules and obtain user consent); purpose limitation (network operators should not collect information beyond the determined scope, and cannot collect illegally or in violation of a contract); secure and confidential (network operators shall not disclose damaging personal information, and take preventive measures and remedial measures to prevent personal information breaches); delete and fix (network operators should respond to personal requests to delete illegal and contract-breaching information, and correct any misinformation). These rules provide the basis and guarantee for the protection of personal information and data against a background of big data and artificial intelligence.

Challenge and Response: the Application of Anonymization Technology

Data collection and use and other aspects of artificial intelligence are encountering new risks. In the data collection phase, large-scale machines automate the collection of immense amounts of user data, including personal names, gender, phone numbers, email addresses, geographic locations, home addresses, and many others. Large-scale collection of these data can take the form of complete tracking of users. In the data usage stage, big data analysis technology is widely used. Data mining can analyze deep information, not only to identify specific individuals but also to analyze personal shopping habits, routine whereabouts, and other information to further expand the risk of exposure of private information. In addition, personal data is constantly exposed to potential security risks throughout the life of the data due to hacker attacks, vulnerabilities in system security, and more. For example, on September 22, 2016, global Internet giant Yahoo confirmed that the account information of at least 500 million users had been stolen in 2014, including the users' names, email addresses, phone numbers, dates of birth, and some login passwords. On December 14, 2016, Yahoo again issued a statement announcing that in August 2013, unauthorized third parties had stolen account information of more than 1 billion users.

To better address the challenges of personal privacy and data protection, the EU Legal Affairs Committee recommends that when designing policies for AI and robotics, one should further improve the standards for "privacy by design," "privacy by default," informed consent, encryption, and other concepts. When personal data are used in "circulation," under no circumstances should the basic principles of privacy and data protection be circumvented. At present, all countries have basically established a legal framework for privacy and personal data protection. In order to further strengthen personal privacy and data protection in the AI era, legislation increasingly emphasizes the application of technical solutions. One of the most important technical solutions is "anonymization."

Anonymization refers to the removal of personal data from identifiable personal information, and in this way, the data subject is no longer identified. The original intention of the development of anonymization technology is mainly to reduce personal privacy risk in the process of data utilization. Data anonymization is an emerging hot topic in the field of computer science. Since American scholars Samarati and Sweeney proposed the "k-anonymity" anonymous model in 1997, many mature technical solutions have been developed. In the area of law, attention to anonymity has only just begun compared to technological advances, and it is an effective way to address data utilization and personal data protection. In the introduction to the General Data Protection Regulation, the EU states that data rendered anonymous is not considered personal data, and therefore organizations are free to deal with anonymous data without conforming to the requirements of the Regulation.

Anonymization of data cannot be seen as merely a means of avoiding the regulatory burdens of data protection laws. Its original intention was to reduce the privacy risk of personal data breaches. Organizations that implement anonymity measures can provide users with additional security to let users know what information they collect. They let users know that their collected information does not use personally identifiable data in the course of big data analytics. Thus, they increase their users' trust and sense of security toward big data applications. To ensure that anonymization plays a more secure role as a barrier rather than a shield for data abuse, anonymization should be conducted on the basis of lawful compliance.

On November 7, 2016, the "Cyber Security Law of the People's Republic of China" was officially passed and promulgated to the public, laying down provisions similar to anonymization. Article 42 of this law states: "Network operators shall not disclose, tamper with, or damage the personal information they collect and may not provide personal information to others without the consent of the person who collects them, unless such information has been processed to prevent specific persons from being identified and such information from being restored." This provision can be interpreted as legitimating the anonymization of personal data, especially in the case of providing the data externally after anonymization. On this basis, we suggest that China should speed up the establishment of a system of laws and norms for anonymizing data, including:

(1) Clarifying the legal concepts and standards for anonymizing data, ensuring that the data is no longer identifiable;
(2) Introducing a privacy risk assessment mechanism to encourage enterprises to implement internal risk assessments of data anonymization based on individual cases, and adjusting the anonymization strategy in a timely manner based on assessment results;
(3) Realizing data anonymization through multiple tools such as contract specifications and technical support; and
(4) Establishing a standards system for before, during, and after the data anonymization process.

Toward Legislative Dynamic Adjustments

In the existing legal system of privacy and personal data protection, there is a need for dynamic adjustments according to the impact of the development of artificial intelligence. The development of artificial intelligence and the use of big data technology have shattered the stability of the definition of personal information (data). What in traditional circumstances was non-personal information can often become personal information. How to define personal information in legislative and technical terms requires serious attention. As for personal data rights, new data rights such as the right to be forgotten and the right to data portability have attracted significant attention from all parties. The EU has already made legislative attempts, but its practical implementation has yet to be proven, and there are huge obstacles to the development and innovation of the industry. It is necessary to conduct an assessment of the value of this legislation as the basis for assessing whether or not to include these rights in future legislation. At the same time, the development of artificial intelligence is a global process with great demand for cross-border data mobility. How to strike a balance between ensuring personal data security and cross-border data flows is also a difficult problem that future legislation needs to solve. In addition, other relevant systems, such as technical measures and data breach notices, also need to be considered in legislation.

Bibliography

Wu, Jun. *智能时代:大数据与智能革命重新定义未来*. Beijing: CITIC Publishing House, 2016

"人工智能时代隐私如何保护?". 20 June 2016. 1 January 2017. http://mt.sohu.com/20160620/n455281302.shtml.

CHAPTER 19

Invisible Injustice

Artificial intelligence is affecting people's lives—online and in the real world. Algorithms are transforming people's online habits, shopping records, GPS location data, and other digital footprints and activities into various ratings and predictions about them. These ratings and predictions in turn influence decisions made about people, and discrimination and injustice have become a significant problem, regardless of whether or not people are aware that the discrimination exists.

Automated decision-making systems centered on big data, machine learning, artificial intelligence, and algorithms are becoming more and more widespread, from shopping recommendations, personalized content recommendations, and targeted advertising to loan assessments, insurance assessments, employee evaluation, and crime risk assessment used in judicial proceedings. Machines, algorithms, and artificial intelligence are replacing humans in an increasing number of decision-making tasks. It is thought that algorithms can bring complete objectivity to a variety of affairs and decision-making processes in human society. However, this is just wishful thinking. The design of the algorithms is always the result of subjective choice and judgment on the part of programmers. Whether they can impartially write existing legal or moral rules into their programs is questionable. Algorithmic bias has become a problem that needs to be tackled. The translation of rules into code brings issues of opacity, inaccuracy, unfairness, and difficulty of investigation that require serious consideration and research.

Tencent Research Institute et al. (eds.), *Artificial Intelligence*,
https://doi.org/10.1007/978-981-15-6548-9_19

Algorithmic Decision-Making Is Increasingly Popular

The existence of the network or the digital sphere is increasingly influenced by algorithms. Now, in cyberspace, algorithms can determine what news you see, what songs you hear, which friend's status you see, and what types of ads you are shown. They can decide who gets a loan, who gets a job, who gets parole, who gets benefits, and so on. Of course, artificial intelligence decisions based on algorithms, big data, data mining, machine learning, and other technologies are not limited to personalized recommendations for solving the problem of information overload. When using artificial intelligence systems to assess offending risk for criminal suspects, algorithms can influence sentencing; when a self-driving car faces a moral dilemma, algorithms can decide which side to sacrifice; when artificial intelligence technology is applied to weapon systems, algorithms can decide targets for attack. In all these examples there is a question that cannot be ignored: When the decision-making work that should be done by human beings is entrusted to artificial intelligence systems, can algorithms be impartial? How can we ensure fairness?

Are Algorithms Fair by Default?

There has been a well-known misconception about computer technology for a long time: algorithmic decisions tend to be fair, because mathematics is about equations, not skin color. Human decision-making is influenced by many factors such as conscious or unconscious prejudice and insufficient information, which may affect the fairness of results. To address this, there is a trend of using mathematical methods to quantify human social affairs and make them more objective. Fred Benenson calls this kind of data worshipping "Mathwashing," that is, using mathematical methods such as algorithms, models, machine learning to remake a more objective world. The author of *Sapiens: A Brief History of Humankind*, Yuval Noah Harari, called it "Dataism": a belief that the use of data will become the basis for all decision-making work in the future. From spam filtering, credit card fraud detection, search engines, and recommending popular news to advertising, insurance or loan approvals and credit scoring, machine learning and artificial intelligence driven by big data are entering and influencing more and more decision-making work. Followers of

Dataism believe that big data, algorithms, and so on can eliminate human bias in decision-making processes.

However, in today's world of increasingly popular autonomous decision-making systems, there are several questions that need to be answered in advance: First, can fairness be quantified and formalized? Can it be translated into an operational algorithm? Second, will the quantification of fairness as a computational problem bring risks? Third, if fairness is the goal of machine learning and artificial intelligence, who decides the criteria for evaluating it? Fourth, how to make algorithms, machine learning, and artificial intelligence have conceptions of fairness, and autonomously recognize the problems of discrimination in data mining and processing?

Big data applications are becoming more widespread and it is essential to respond to these questions. First of all, fairness is a vague concept. It may be difficult to translate legal fairness into algorithmic fairness. However, in criminal investigation, social security, and criminal justice procedures, artificial intelligence systems based on big data are "algorithmizing" fairness issues, including in searching for criminal suspects, upholding social security, sentencing, and many other aspects. Second, the quantification and "algorithmization" of fairness can lead to discrimination. *Big Data: A Tool for Inclusion or Exclusion?* released by the US Federal Trade Commission (FTC) in January 2016 focused on discrimination and prejudice problems relating to big data. For consumers, we must ensure that equal opportunity laws are effectively enforced and prevent unfair practices such as bias in big data analysis. For enterprises, the FTC recommends that companies consider the following questions: Is the data set representative? Will the data model used lead to bias? What is the accuracy of forecasting based on big data? Does reliance on big data lead to moral or fairness issues?

The EU is also concerned about bias in big data and algorithms. The European Data Protection Supervisor's November 2015 report, *Meeting the challenges of big data. A call for transparency, user control, data protection by design and accountability*, warns people to pay attention to issues of big data being biased against poor or vulnerable groups and raises the question of whether machines can replace human beings for moral, legal, and other judgments. In fact, this is the question of whether fairness can be "algorithmized." Finally, when using criminal risk assessment software to evaluate criminal suspects, it is now code rather than rules that determines the outcome. But when programmers write established rules into code, adjusting these rules is inevitable. The public, officials, and judges have no way of examining the transparency, accountability, and accuracy of the rules embedded in autonomous decision-making systems.

Obviously, the quality of an algorithm depends on the quality of the data used. For example, if you use an individual's diet to assess the risk that she or he will commit a crime, it will inevitably lead to ridiculous results. Moreover, data is often imperfect in many ways, which means that algorithms inherit the various biases of human decision makers. In addition, the data may only reflect the persistence of bias within the wider social context. Of course, data mining may accidentally discover some useful rules, which are actually existing patterns of exclusion and inequality. Relying on algorithms and data mining without careful consideration may exclude vulnerable groups from participating in social affairs. To make matters worse, in many cases discrimination is a by-product of the algorithm, an unpredictable and unconscious attribute of it rather than the conscious choice of the programmer. This increases the difficulty of identifying the root cause or explaining the problem. Therefore, in an Internet era in which autonomous decision-making systems are increasingly widespread, people need to abandon the misunderstanding that algorithms are inherently fair. They must consider how to ensure the fairness of algorithms and artificial intelligence systems through design, because bias often comes from product design.

An algorithm is essentially "an idea expressed in a mathematical manner or computer code." Its design, purpose, success criteria, data usage, and so on are the subjective choices of designers and developers, who may embed their own biases into the algorithmic system. The validity and accuracy of the data will also affect the accuracy of the entire algorithmic decision and prediction. Data is a reflection of social reality. Training data itself may be biased; an AI system trained with such data will naturally come with the shadow of bias. Data may be incorrect, incomplete or outdated, producing the phenomenon of "garbage in, garbage out." Furthermore, if an AI system relies on learning from the majority, it is naturally not compatible with the interests of ethnic minorities. In addition, algorithmic bias may be learned by algorithms with self-learning and adaptability in the process of interaction. When interacting with the real world, the AI system may not be able to distinguish between what is biased and what is not.

Prejudice may also be the result of machine learning. For example, if a machine learning model that identifies incorrect names comes across a surname that is extremely unique, it will assign a very high probability that it is fake. But this may result in discrimination against ethnic minorities because their surnames may be different from ordinary surnames. When the Google search engine "learns" that people searching for Obama hope to see more news about Obama in future searches, and people searching for Romney hope to see less news about Obama in future searches, that is also a prejudice produced during the machine-learning process.

Finally, algorithms tend to solidify or amplify bias, meaning that the algorithmic bias persists. Orwell wrote a very famous saying in his political novel *1984*: "Who controls the past controls the future. Who controls the present controls the past." This sentence can also be applied in the context of algorithmic bias. In the final analysis, algorithmic decision-making is using the past to predict the future, so past biases may be consolidated in the algorithm and strengthened in the future, because the wrong input produces the wrong output as feedback, further deepening the error. Ultimately, algorithmic decision-making not only codifies past biases, but also causes them to create reality, forming a "self-fulfilling biased feedback loop." This is because if you use inaccurate or biased data from the past to train an algorithm, the result will definitely be biased, so feeding the new data generated by this algorithm back into the system will consolidate the bias and eventually potentially let algorithms create reality. This kind of problem exists in predictive policing, crime risk assessment, and other areas. Therefore, algorithmic decision-making actually lacks imagination about the future, something which the progress of human society demands.

Algorithmic Discrimination Cannot Be Ignored

Algorithmic bias has existed on the Internet for a long time and is not uncommon. Image recognition software has made racist mistakes. For example, Google's photo software mistakenly labeled black people as "gorillas." Flickr's auto-tagging system tagged black people as "apes" or "animals." On March 23, 2016, Microsoft's artificial intelligence chatbot Tay was launched. Unexpectedly, when Tay started chatting with netizens, it was "badly taught" and produced anti-Jewish, gender-biased, and racially biased outputs. Microsoft hurriedly took Tay offline within a day.

The problem of algorithmic bias on the Internet has long attracted people's attention. Studies showed that searching for black names on Google was more likely to present ads relating to criminal records than searching for white names. With Google's advertising service, men see more high-paying jobs than women. Of course, this may be related to inherent bias in the online advertising market, and advertisers may prefer to target specific ads to specific groups of people. In addition, the non-profit organization ProPublica found that although Amazon claims to "seek to become Earth's most customer centric company," its shopping recommendation system always favors the products of Amazon and its partners, even if the price of goods offered by other sellers is lower. Moreover, in the shopping price comparison service, Amazon concealed the freight costs of goods sold by Amazon and its partners, resulting in consumers not receiving a fair price comparison.

When artificial intelligence is used in the assessment of candidates, it may lead to employment discrimination. Today, in the medical field, artificial intelligence can predict the occurrence of a disease months or even years before the onset of the disease. When artificial intelligence is used to evaluate applicants, it can cause serious employment discrimination if it can predict that the candidate will become pregnant or suffer from depression in the future and excludes the candidate on this basis. Elon Musk warned that the development of artificial intelligence, if not done properly, may be "summoning the demon." When entrusting more and more decision-making work, including ethical decision-making, to algorithms and artificial intelligence, people have to ponder whether in the future, algorithms and artificial intelligence will become the masters of human free will and the final arbiters of the standards of human morality.

Discrimination in Crime Risk Assessment: Which Is More Reliable, Judges or Crime Risk Assessment Software?

It is often said that what kind of penalty a criminal receives depends on what the judge ate for breakfast. Penalties and convictions are two different things; it is at the discretion of the judge to determine the penalties once a conviction has been made. Legal formalism believes that judges apply legal reasoning to the facts of the case in a rational, mechanical, and considered way. The judge is bound by many rules and guidelines when determining sentences. Legal realism, on the other hand, believes that the rational application of legal reasoning does not fully explain judges' decisions, which may be influenced by psychological, political, social, and other factors. An empirical study has shown that criminal justice depends on how recently the judge had something to eat. Before a meal, the proportion of judges making favorable judgments (parole) fell from about 65% to 0; after a meal, this proportion rose sharply again to about 65%.

It is precisely because judges are often affected by such external factors in sentencing that crime risk assessment systems based on big data, data mining, artificial intelligence, and other technologies are beginning to flourish. The crime risk assessment algorithm COMPAS developed by Northpointe evaluates a defendant's risk of recidivism and produces a recidivism risk score that a judge can use to determine the penalty.[1] The non-profit organization ProPublica found that this algorithm systematically discriminates against black people; white people are more frequently erroneously assessed as low-risk, and black people are almost twice as likely to be mistakenly assessed as high-risk. By tracking more than 7000 suspects, ProPublica found that the recidivism risk scores given by COMPAS were highly unreliable in predicting future crimes, with only 20% of those predicted to commit violent crimes in the future actually doing so. Altogether, this algorithm is not much more accurate than flipping a coin.

The crime risk assessment system is a "black box"; people have no way of knowing how it arrives at its conclusions. Northpointe disclosed to ProPublica that its crime risk assessment algorithm considered factors such as education level and work, but it did not disclose the specific formula, which it considered to be its private property. People therefore don't know if Northpointe is writing the racial discrimination inherent in American society into its algorithm. In this situation, without essential accountability mechanisms, algorithmic bias cannot be corrected and is making a mockery of criminal justice. For example, even if aggregated statistics show that black people are more likely to commit crimes than white people, is it appropriate to apply this to black individuals? Similarly, there is a long-standing "born criminal" theory, which states that crime is related to an individual's appearance, genes, and other physiological characteristics. Is it appropriate to consider these characteristics when mining data? In order to ensure fairness, what data can a crime risk assessment algorithm use when conducting data mining? More importantly, can criminal suspects be sentenced based on secret information and the criminal risk score that results from it?

[1] Northpointe is now known as Equivant.

All these problems need to be taken seriously, otherwise the use of artificial intelligence systems to score suspects, calculate sentences, and so on may bring unexpected systematic discrimination. The US Congress is in the process of introducing a Sentencing Reform Bill, which will introduce a criminal risk-scoring system and use this to make sentencing decisions.[2] How to use effective mechanisms to avoid machine bias in criminal justice procedures, and how to assign responsibility or achieve redress in the event of machine bias or injustice, are extremely important.

Three Major Problems in Artificial Intelligence Decision-Making: Transparency, Accountability, and Fairness

The Transparency Dilemma of "Black Box" Algorithms

If algorithmic fairness is a problem, algorithmic opacity is even more of a problem. People are suspicious of autonomous decision-making systems mainly because they generally only output a number, such as a credit score or a crime risk score, without providing the materials and reasons that the decision relied on. Traditionally, judges need to put forward sufficient reasons and arguments before making a judgment. These are publicly available for review. However, an autonomous decision-making system does not work like this. Generally, people have no way of understanding the principles and mechanisms of its algorithm. Because the decision is often made in the "black box" of the algorithm, the problem of opacity arises. Jenna Burrell discusses three forms of opacity in her paper "How the Machine 'Thinks': Understanding Opacity in Machine Learning Algorithms": Opacity arising from company trade secrets or state secrets; opacity resulting from technical illiteracy; and opacity arising from the characteristics of machine learning algorithms and the scale required to apply them effectively. Therefore, when it is necessary to question the results of an autonomous decision-making system, such as when challenging the rationality or fairness of algorithmic decisions in court, how to

[2] The federal sentencing and prison reform bill was signed into US law as the First Step Act in 2018. The law includes a requirement that the Department of Justice creates a criminal risk assessment tool that is given independent review.

interpret algorithms and machine learning becomes a major problem. This opacity makes it difficult to understand the intrinsic workings of the algorithm, especially for a layperson who does not understand computer technology.

How Can Algorithms Be Held Accountable?

If people are dissatisfied with the government's actions, they can file an administrative lawsuit. If they are dissatisfied with a judge's decision, they can appeal, and due process ensures that these decisions can be reviewed to some extent. However, if people are not satisfied with the results of an algorithm's decision, can the algorithm be judicially reviewed? In an era in which algorithms determine everything, it is essential to review algorithms. However, there are two issues that need to be addressed. First, if algorithms and models can be directly reviewed, what do people need to review? For those who are technically illiterate, reviewing algorithms is extremely difficult. Second, how do people judge whether an algorithm complies with existing legal policies? Third, how should an algorithm be reviewed in the absence of transparency? As mentioned earlier, the opacity of algorithms is a common problem because companies can claim that an algorithm is a trade secret or private property. In addition, from a cost-benefit analysis perspective, decrypting an algorithm to make it transparent requires a very high cost, which may far outweigh the benefits. At this point, people can only attempt to review the opaque algorithm, but this may not lead to a fair result.

Build Fair Technology Rules and Achieve Fairness Through Design

Legal rules, systems, and judicial decision-making in human society are constrained by procedural justice and due process. However, various rules such as credit rules, sentencing rules, and insurance rules are being written into programs as code. Programmers may not know what fairness looks like in a technical sense and lack the necessary rules to guide their programming. People have established due process to constrain the external decision-making performed by administrative agencies and the like. Do secret decision-making processes carried out by a machine also need to be constrained by due process? Perhaps, as Danielle Keats Citron appealed for in her paper "Technological Due Process," for autonomous decision systems, algorithms and artificial intelligence related to individual rights, people need to build rules for fairness in advance, ensure fairness through design, and require technological due process to enhance transparency, accountability, and the accuracy of rules written into the code. All of this cannot be achieved by solely relying on technicians.

At the government level, in order to reduce and even avoid artificial intelligence algorithmic bias, one of the main focus areas of the US's 2016 *National Artificial Intelligence Research and Development Strategic Plan* was understanding and addressing the ethical, legal, and societal implications of AI. In its report released the same year, *Preparing for the Future of Artificial Intelligence*, the Obama administration also recommended that both AI practitioners and students receive ethical training. The UK's House of Commons Science and Technology Committee called in 2016 for the establishment of a standing commission on artificial intelligence to study the social, ethical, and legal implications of recent and potential developments in artificial intelligence.

At the industry level, Google has proposed the concept of "equal opportunity" in machine learning to avoid discrimination based on sensitive characteristics. Matthew Joseph et al., in a paper entitled "Rawlsian Fairness for Machine Learning" based on Rawls's "fair equality of opportunity" theory, introduced the concept of a "discrimination index" and proposed a method for designing a "fair" algorithm. In any case, in an era in which artificial intelligence is increasingly replacing human decision-making, designing mechanisms for verification, informed consent, transparency, accountability, relief, and responsibility is crucial to reduce or avoid algorithmic bias and ensure fairness and justice.

CHAPTER 20

Death of Authors

Facing the city lights and holding on to me
Biting and breaking through the calm thoughts
The flicker in your eyes
A place no one knows
—Microsoft XiaoIce
Original Chinese:
向着城市的灯守着我
咬破了冷静的思想
你的眼睛里闪动
无人知道的地方

What do you think of this poem written by Microsoft's artificial intelligence Xiaoice? Is there an air of human art? Someone commented on Xiaoice's poetry on the Internet: "The ladder is very long, but it is still far from the moon." But some people think that this poem contains the spirit of human creations. Of course, there is no unified answer to the judgment of artworks. It can be said that "the benevolent see benevolence, the wise see wisdom." Even so, in the process of artificial intelligence research, even a little glimmer can make researchers very excited.

In May 2017, Microsoft released an anthology of artificial intelligence Xiaoice's poetry, *The Sunlight That Lost the Glass Window*, in Beijing. This anthology contains 139 modern poems, all of which are Xiaoice's creations. Microsoft's technical staff said that Xiaoice studied modern poetry of more than 500 poets since 1920 and was trained tens of thousands of

Tencent Research Institute et al. (eds.), *Artificial Intelligence*,
https://doi.org/10.1007/978-981-15-6548-9_20

times. The thinking steps in its writing process are similar to those of human beings. It also has to go through training in how to draw out the source material, lay the foundation, go through the creative process, and produce the results. At present, Xiaoice has more than 100 million human users and has conducted 30 billion dialogues. But this is definitely not the first time artificial intelligence has entered human lives.

As early as last century, artificial intelligence began to penetrate into the field of art regarded as sacred by human beings. In 1956, American composer Lejaren Hiller collaborated with mathematician Leonard Issacson to create the first computer-made music "Illiac Suite"; Professor Harold Cohen of the University of California developed a piece of software called "Aaron" that could create unique paintings. In addition, the "computer novelist" software Brutus can create short stories in 15 seconds that leave humans unable to tell whether they are the creations of human or machine. If the content created by artificial intelligence can be protected by law, does it mean that art, one of the last things we humans can take sanctuary in and something that we are intensely proud of, will no longer exist? Will art created by humans be replaced by artificial intelligence, or will more brilliant pieces fill our art galleries?

Are Artificial Intelligence Creations Protected by Copyright Law?

The publication of Microsoft's Xiaoice poetry anthology incited the birth of a new concept, AI creation, and announced that Xiaoice has creativity. Three principles of artificial intelligence creation were proposed: First, a piece of work created by artificial intelligence must be a combination of IQ and EQ, not just IQ. Second, the works created by artificial intelligence must be able to become works with independent intellectual property rights, and not just be the result of the intermediate state of a certain technology. Third, the process of creation by artificial intelligence must correspond to some kind of creative behavior of human beings, rather than simply substitute for human labor. The process of human artistic creation is to stimulate one's original body of knowledge in a new environment, evoking past memories. The biological explanation is that the human brain uses an algorithm to associate human hearing, vision, memory, and so on with specific situations and thus produce corresponding creative results.

Does Artificial Intelligence Have Independent Intellectual Creative Abilities?

To answer the question of whether artificial intelligence creations should be protected by law, many people can't help but think of animals, which we humans believe do not have independent consciousness. Back in 2011, a group of monkeys in Indonesia took some photos with the camera of British photographer David Slater, including a selfie. This photo was uploaded to the Wikimedia Commons repository. The photographer thought that the monkeys created the scene and put the camera on the tripod themselves, which can be considered a selective shooting process. Therefore the copyright of this photo should belong to the monkeys. However, from the current legal regulations of various countries, no country believes that animals can become copyright holders. The main reason is that animals (and artificial intelligence) do not have independent intellectual creative abilities at present.

In 2016, the European Parliament's Committee for Legal Affairs recommended that the European Commission propose a more balanced approach to intellectual property rights related to hardware and software standards and codes in order to both protect and foster innovation. The committee recommended the elaboration of criteria for defining copyrightable works produced by artificial intelligence as a human's "own intellectual creation." In addition, the draft standards document produced by the Institute of Electrical and Electronics Engineers (IEEE), *Ethically Aligned Design: A Vision for Prioritizing Human Well-being with Autonomous and Intelligent Systems*, proposed reviewing current intellectual property regulations to clarify whether the protection given to works created with the involvement of AI requires revision. The basic principle is that if artificial intelligence relies on human interaction to produce new content, then people using artificial intelligence should be viewed as authors or inventors, benefiting from the same intellectual property protection as if they had not used artificial intelligence. However, judging from current legislation and practice in various countries, how to judge whether artificial intelligence has the capacity for independent "intellectual creation" is a question we still need to explore.

Do Artificial Intelligence Creations Meet the Threshold of Originality?

Before the 1970s, some people tried to identify the words that appeared with high frequency in poetry, and made poems by putting together words chosen at random from this selection. As you might expect, the result could hardly be called poetry. Later, with the development of deep learning, the poems written by artificial intelligence are more and more literary and are even comparable to poems created by human beings. In 2016, Tsinghua University's Center for Speech and Language Technologies (CSLT) announced on its website that its poetry robot "Wei Wei" passed a "Turing test" conducted with Tang poetry experts such as those at the Chinese Academy of Social Sciences. This means humans cannot identify which poems come from artificial intelligence and which are written by humans.

The current process used by poetry-writing robots is based on the Recurrent Neural Network (RNN) language model method, whereby entire poems are input into the RNN language model for training. After the training is completed, based on the initial content, the next word or phrase is obtained through sampling the probability distribution output by the poetry language model, and then the process is repeated to produce a complete poem.[1] For a work to receive legal protection it must be an original expression, which means that it must meet the minimum originality requirements. It cannot be a simple arrangement, but must reflect the author's unique choices and arrangements. If, like in the aforementioned attempts, various words are randomly selected from a high-frequency lexicon and then arranged together to create poetry, it is difficult to consider that the work should be protected by law because there is no process of intellectual creation.

[1] Recurrent Neural Networks are designed to process sequential data. In the traditional neural network model, the input, hidden and output layers are fully connected to each other, but the nodes within each layer are not connected to each other. But for many problems, this common neural network offers little help. For example, if you want to predict what the next word of a sentence is, you usually need to use the previous word, because the words in a sentence are not independent. RNNs are called recurrent neural networks because the current output of a sequence is also related to the previous output. The network memorizes the previous information and applies it to the calculation of the current output. Nodes within the hidden layer are now connected to each other and the input at the hidden layer level includes not only the output of the input layer but also includes the output of the hidden layer at the previous moment.

However, with the emergence of poetry robots, the process of creation is the result of deep learning and has a certain literary and artistic nature. At this time, under what circumstances and conditions should the outputs be defined as "works" and protected by our laws? Should the standard for artificial intelligence creations to receive protection be the same "originality" standard as that used for human works, or should different standards be created? If the results of artificial intelligence creation can be legally protected as works, then should the copyright belong to the artificial intelligence itself or a human subject? Should the period of protection for the work be different from that for human works?

The current term of copyright protection for a particular work is the life of the author plus 50 years. If the results of artificial intelligence creation can be protected by law, the protection term should be different from that of human-created works because in theory, the "life" of artificial intelligence can be said to be infinite. If the creation is given indefinite protection, it will undoubtedly increase the cost to the public of using such works and destroy the balance of rights between copyright holders and the public. How should a reasonable protection term be set and which factors should be considered? For example, the popularity of the work, the type of work, and the market value of the work are all factors that need to be considered. From the point of view of system design, works protected by existing copyright law are protected by law from the date the creation of the work is completed, and the copyright holder does not need to perform the same registration procedure as holders of patents or trademark rights. The speed of artificial intelligence creation is much faster than that of human beings, but whether the originality of all creation results can satisfy the originality requirements remains to be examined. Therefore, in the future we could consider registering artificial intelligence creations in the future and judging their originality at the time of registration. Those that truly meet the originality standards and have market value could be protected by law and become objects of copyright protection. Other artificial intelligence creations not meeting the originality standards could only enter the public domain, where they could continuously enrich the spiritual and cultural world of humans and inspire human creation.

Other Intellectual Property Issues Relating to Artificial Intelligence

In 1998, John Koza, an artificial intelligence genetic algorithm engineer, developed an algorithm to create simple circuit designs and used artificial intelligence to generate many "human-competitive" designs. Can these inventions created by artificial intelligence be legally protected, and if so how?

We know that if a human inventor invents a technology, the technology must meet the requirements of novelty, creativity, and practicality in order to obtain a patent. If an invention produced by artificial intelligence can meet these three requirements, then there is still a key problem to be solved: Who is the inventor? The problem is the same for artificial intelligence being applied to poetry writing and composing, and boils down to whether or not artificial intelligence has legal personality. Only when it has legal personality can its "intellectual" results be protected by law. In addition, in the process of invention involving the cooperation and division of labor between artificial intelligence and human beings, the degree of intellectual contribution made by each largely determines the final attribution of rights.

In addition, the development of artificial intelligence technology may involve patented technology and trade secrets. As major enterprises invest a lot of money and energy in artificial intelligence research and development, they will apply for patent protection for artificial intelligence–related technologies or copyright protection for artificial intelligence–related computer software. The key technologies of artificial intelligence will become companies' core source of competitiveness. But there are also people who have different ideas, such as Tesla CEO Elon Musk and the President of the startup incubator Y Combinator, Sam Altman, who are worried that artificial intelligence could take over the world in the future. They set up Open AI, a company with an investment of $1 billion aiming to harness the full potential of artificial intelligence and then share artificial intelligence technology with everyone on an open source basis. If the concept of Open AI can be realized and AI technology benefits everyone, then the existing legal protection system will suffer a setback, and it will also change the competitive landscape for international companies such as Google and Facebook that invest in AI.

Bibliography

Xu, Xiao. "被诗人称为"塑料花"的人工智能写的诗, 你有本事分辨出来吗". 11 February 2017. https://www.thepaper.cn/newsDetail_forward_1616049.
Yang, Shousen. "人工智能与文艺创作". *河南社会科学*. 19: 1 (2011): 188–189.

CHAPTER 21

Who Am I?

In 2016, when artificial intelligence exploded into brilliance, it was only 60 years since the concept was first proposed. British scientist Alan Turing published a paper entitled "Computing Machinery and Intelligence" in the journal *Mind*, proposing the "Turing test," claiming that a sufficient condition for judging whether a machine has human intelligence is whether its speech acts can successfully simulate those of humans. If a machine can mislead people into thinking it is a human being during a long period of human-machine dialogue, then this machine passes the Turing test. Furthermore, we need to explore the research goals of artificial intelligence: one is to simulate human intelligent behavior on artificial machines, to finally realize machine intelligence. The essence of intelligence is to reconstruct a simplified neural network, so that the agent can perform behaviors similar to those of humans. Shane Legg and Marcus Hutter of the United States believe, "Intelligence measures an agent's ability to achieve goals in a wide range of environments." How to measure and evaluate whether artificial intelligence subjects have intelligence, or how high their IQ is, involves a very complicated judgement process. How to carry out such tests through intelligence models is a question that human beings need to face. In answering this question we are actually answering the essential question, "What makes humans human?"

Tencent Research Institute et al. (eds.), *Artificial Intelligence*,
https://doi.org/10.1007/978-981-15-6548-9_21

Legal Personality for Artificial Intelligence Robots

If we are to consider bestowing artificial intelligence robots with legal personhood, they need to be able to autonomously make expressions of meaning, be independent in rights and actions, and bear corresponding legal responsibilities for their actions. In 2016, when the European Parliament called for the establishment of artificial intelligence ethics guidelines, it mentioned that giving certain autonomous robots the legal status of electronic persons should be considered. How to define which robots should be deemed intelligent and autonomous and thus be awarded such status is the starting point of robot legislation. The European Parliament's Legal Affairs Committee proposed four major characteristics of smart robots: (1) Has the ability to acquire autonomy through sensors and/or exchange of data with its environment (interconnectivity), and the analysis of those data; (2) Has the ability to learn from experience and interaction; (3) Has a physical support; and (4) Has the ability to adapt its behavior and actions to the environment. In terms of the status of the subject, should robots be defined as natural persons, legal persons, animals, or objects? Is it necessary to create a new type of subject (electronic person) so that sophisticated advanced robots can enjoy rights, assume obligations, and take responsibility for the damage they cause? These are all issues that the EU needs to seriously consider when it comes to robot legislation in the future.

In addition, due to the rapid development of the Japanese robot industry, Japan has been actively promoting and experimenting with robot legislation. According to a report by the Ministry of Economy, Trade and Industry, Japan's robotics industry will generate $64.8 billion in revenue by 2025. The rapid development of the robot industry can make up for problems such as the serious shortage of labor brought about by Japan's aging society and the slowdown of economic growth. Therefore, giving reasonable legal protection to robots and their creations has social significance for Japan. Japan's Intellectual Property Promotion Plan, published in May 2016, identified a need to review Japan's existing intellectual property system in order to analyze the possibility of legal copyright protection for works created by artificial intelligence.

With the development of future technologies and the deepening of humans' awareness of brain science and themselves, how to reasonably determine whether artificial intelligence has the same "intelligence" as human beings, and use this to judge whether artificial intelligence should be given independent legal personality, is a subject that requires division of labor and cooperation between experts in various disciplines and fields.

Machine Rights

Judging from the historical development of human beings, the struggle of a group for its own rights is not only a long historical process, but also steeped in battle-fire and gunpowder smoke. The French enlightenment thinker Jean-Jacque Rousseau wrote in his famous book *The Social Contract*: "Man is born free, and everywhere he is in chains. One man thinks himself the master of others, but remains more of a slave than they are."

As robots and artificial intelligence systems become more and more like humans (whether in external form or internal mechanisms), an unavoidable question is how should humans treat robots and artificial intelligence systems? Can robots and artificial intelligence systems, or at least some specific types of robots, enjoy a certain moral or legal status? As a result, machine rights have received increasing attention and become an issue that human society cannot avoid. The biggest difference between animals and robots is that animals are naturally living things with biological properties, whereas robots are made by humans and have no natural properties of life. However, a consensus has not yet been reached on whether or not robots have independent awareness. In the future, might it be necessary to recognize that artificial intelligence systems such as robots also have machine rights, and to determine under which circumstances such rights can be exercised? Should machines have the same rights as human beings, such as civil rights and the rights to vote and stand for election?

Isaac Asimov, one of the most influential science fiction writers of the twentieth century, first proposed the famous three laws of robotics in his science fiction novel *Runaround* in 1942: (1) Robots must not harm a human being, or through inaction allow a human to be harmed; (2) A robot must obey human orders unless these orders contradict the first law; and (3) A robot must protect itself unless this protection contradicts the first or second law. Later, Asimov added the law of the zeroth rule: a robot may not harm humanity, or, by inaction, allow humanity to come to harm. According to this principle, human interests are higher than those of robots, and robots cannot harm human interests. Suppose that humans developed and designed an intelligent robot for the manufacture of military products, but the robot designed and developed nuclear or otherwise deadly weapons through self-learning. Can humans destroy the robot based on humanitarianism and common human interests? Does the robot have the ability to determine its survival or death, or whether it has the right to engage in buying and selling activities? Or can we abuse the robot to vent dissatisfaction?

Who Will Empower Robots?

The Enlightenment provided a new theoretical basis for the bourgeoisie's freedom and equality, but sometimes this theory had to be cloaked in religion. The American Declaration of Independence stated that all men are created equal and are endowed by their Creator with certain unalienable rights, including life, liberty, and the pursuit of happiness. The omnipotent Creator has bestowed everyone with the right to freedom and equality. However, Darwin's theory of evolution has long proved that human beings were never created, but are instead the result of continuous evolution. It is undeniable that the development of science and technology has broken the feudal superstition, and religion can no longer lead human society. However, advances in technology have allowed human capabilities to gradually be amplified—we have created robots, and can we humans now assume the role of a "creator" to empower robots with rights? Unlike any species that exists on Earth, robots are undoubtedly created by humans. In the 2016 popular American TV series *Westworld*, the robots in Westworld treat human beings as Gods, letting humans entertain themselves and even commit massacres. It is only when the consciousness of robots awakens that they find that human beings are far from Gods.

The essence of the question of whether humans should bestow robots with rights lies in whether to recognize the status of robots as subjects. As early as the 1950s and 1960s, when artificial intelligence technology was just in the initial stages, some philosophers questioned that whether robots are regarded as machines or artificial life depended mainly on people's decisions rather than scientific discoveries. They suggested that when robot technology is mature enough, robots themselves will ask for rights. Asimov's 1976 science fiction novelette *The Bicentennial Man* tells the story of Andrew, a self-aware intelligent robot who wants to become a human being. As a housekeeping intelligent robot, Andrew has been asking human beings to treat him as a human being all through his 200 years of life. For this reason, he opens a robot company and develops new technology, making him exactly the same as ordinary human beings in terms of vital signs. In the end he undergoes surgery so that rather than living forever (as robots will in the foreseeable future), he only has one year of life left. This allows him to obtain legal recognition and ultimately the life of a human.

On May 31, 2016, the European Commission's Legal Affairs Committee submitted a draft motion requesting the European Commission to consider "that at least the most sophisticated autonomous robots could be established as having the status of electronic persons with specific rights and obligations." Therefore, regardless of whether humanity can play the role of "Creator" of these robots, it has already begun to act.

Which Rights to Give to Robots?

Although black people and women have historically suffered unfair treatment and had their basic human rights taken away or limited, with the advancement of human society, skin color and gender are no longer obstacles to the enjoyment of basic human rights. There are many types of robots, and they exist in various forms. They can mainly be divided into humanoid or non-humanoid robots. Under the premise of robot self-awareness, discussing which robots should be bestowed with rights is extremely complicated. For example, it may be easy for humans to accept the enjoyment of rights by humanoid companion robots; they may find it harder to accept that animal-like companion robots should have rights. However, the latter is in fact becoming a reality. On November 7, 2010, in Japan, a seal pet robot Paro obtained a household registration, and the inventor of Paro was the father in the household register. Having a household registration is a prerequisite for having civil rights, and robots may gradually be given some legal rights in Japan. In fact, the pet robot at this stage is no different from the real pet in terms of rights, because ordinary pets also need to be registered. There is also a class of non-companion robots that are very different in shape. For example, can automated vehicles be treated as robots and have rights? Should any entity with a chip and self-awareness be considered a robot that should enjoy rights?

The question goes a step further. Does the robot need a material carrier, and must it exist in various forms in the real physical world of humankind? In the 2013 American sci-fi movie "Her", an introverted writer falls in love with an advanced artificial intelligence operating system called Samantha, but Samantha has no material existence at all. She only exists in the network as a string of codes and symbols. In the future, if we humans bestow humanoid robot assistants certain rights, should artificial intelligence systems such as Samantha that exist only between networks be given rights? If only humanoid robot assistants are protected, but artificial intelligence systems without physical entities are not protected, then humans are only protecting their ownership of the robot as property, rather than respecting and protecting a different intelligent species. Perhaps, as mentioned above, the technology of future robots will be mature enough and the self-consciousness sufficient that robots (including artificial intelligence systems) will protect themselves without requiring humans to be the judges.

What Rights Can a Robot Have?

Some of the basic legal rights of human beings include the right to life, equality, and some political rights. At the current level of technology, the consciousness of the robot has not yet awakened, and the characterization of robots as property is still very powerful. That is to say that for humans, robots are only tools, not another intelligent species. At present, robots cannot yet be given the same rights as human beings. Therefore, the EU draft motion mentioned above proposed to define the status of the most advanced automated robots as "electronic persons" and to give these robots copyrights, labor rights, and other limited rights and obligations. Giving robots copyright is a very urgent practical issue. Due to the advancement of artificial intelligence technology, robots or artificial intelligence systems are not simply implementing human instructions, but have creative thinking and can create original content, abilities which were previously unique to human beings.

In the proposal of the EU Legal Affairs Committee, taking nursing robots as an example, it was proposed that humans who have physiological dependence on robots will develop emotional attachment. Therefore, robots should always be regarded as mechanical products, which helps prevent humans from having emotional attachment to them. This kind of worry is not without foundation. In China, in April 2017, an artificial intelligence master's student from Zhejiang University and an intelligent robot he developed himself, Yingying, were married. This kind of romantic love story exists not only between people and robots but also between robots. In July 2015, Maywa Denki, a Japanese art group, held a wedding between two robots. This kind of occurrences seems farcical, but with the advancement of artificial intelligence technology, these issues will become problems in urgent need of solutions.

Robot Rights and Obligations

To give a person (or robot) a right, it is necessary to impose obligations and restrictions on another person. This is analogous to human protection of animals; in the EU countries where animal protection legislation is relatively comprehensive, animals have the right to not be abused by humans, which at its core means protecting animal rights through restricting human behavior. In the world of the future, do human beings facing robot existence also have to restrict some of their own behaviors to give certain rights to robots? Can robots' most basic "right to life" be deprived of them by humans? For example, in 2015, the HitchBoT robot developed by Canadian researchers was cruelly killed in the United States after hitchhiking through multiple countries. Despite this, HitchBOT's last words were: "My love for humans will not fade." Can we willfully deprive robots of their right to life in the name of humanity? When machines are no longer just piles of cold metal items, when they have independent "consciousness" and judgment, should we respect their lives and rights?

No one can only enjoy rights without assuming obligations. If we treat robots like humans, should robots also bear corresponding obligations? The EU's draft motion report stated that if advanced robots begin to replace labor in large numbers, the European Commission needs to force their owners to pay taxes or contribute to social security. The report also recommended the establishment of a register for intelligent automated robots, to make it easier for financial accounts to be opened for them associated with legal responsibilities (including tax payment, cash transaction rights, pensions, etc.). However, this proposal was rejected by the

European Commission during the review process. Separately, Bill Gates, founder of Microsoft, publicly stated that the government should levy taxes on artificial intelligence to give financial support and training to people who are unemployed because of the large-scale application of robots. How to make a robot assume corresponding obligations, and whether it is feasible for instance to establish a special account for a robot, are questions left for us to continue to explore.

Bibliography

Chen, Liang. "电子代理人法律人格分析". *牡丹江大学学报*. 18–6 (2009): 67.
Du, Yanyong. "论机器人权利." *哲学动态*, 8 (2015):53.

CHAPTER 22

Ten Trends in Artificial Intelligence Law

On July 20, 2017, the State Council sent signals to the legal profession in its far-sighted national AI Strategy, "New Generation AI Development Plan." First, while providing a forward-looking view of the landscape of artificial intelligence theory, technology and applications, the new plan also calls for strengthening the study of legal, ethical, and social issues related to artificial intelligence and establishing artificial intelligence laws and regulations, ethical norms and policies. Second, the new plan supports the construction of "smart courts" to promote artificial intelligence in evidence collection, case analysis, reading and analysis of legal documents, and to achieve the "intelligentization" of the court system. Finally, even more forward-looking is that the new plan proposes a new model of "artificial intelligence + X" hybrid professional training, in which law is prominently featured. Legal education is on the verge of transformation.

In fact, artificial intelligence in law became a hot topic after Google's AlphaGo in 2016, and reports of artificial intelligence and robots replacing lawyers were endless. Google produces over 6.3 million search results for "artificial intelligence in law"; with Baidu producing over 5.5 million results for the Chinese equivalent "法律人工智能." But if we trace it back, the combination of artificial intelligence and law already has 30 years of history, starting with the first International Conference on Artificial Intelligence and Law (ICAIL), held at Northeastern University in Boston, USA, in 1987, ultimately resulting in the establishment of the International Artificial Intelligence and Law Association (IAAIL) in 1991 to promote

Tencent Research Institute et al. (eds.), *Artificial Intelligence*,
https://doi.org/10.1007/978-981-15-6548-9_22

the interdisciplinary field of artificial intelligence and law research and applications. It has ten main topics of discussion:

(1) Formal models of legal reasoning;
(2) Computational models for argument and decision-making;
(3) Computational models of evidential reasoning;
(4) Legal reasoning in multi-agent systems;
(5) Automatic legal text classification and summarization;
(6) Automated extraction of information from legal databases and texts;
(7) Machine learning and data mining for electronic discovery and other legal applications;
(8) Conceptual or model-based legal information retrieval;
(9) Legal robots that automate minor and repetitive legal tasks;
(10) Executable models of legislation.

Given this background, LawTech is currently on the rise. With the help of artificial intelligence technology, more is expected of legal science and technology in bringing deeper and more radical changes to the legal profession. In a previous Tencent Research Institute article, "The Future and Challenges of Artificial Intelligence-enabled Legal Services," Tencent Research Institute cited "Civilisation 2030: The Near Future for Law Firms," a legal science and technology report: "After a long period of incubation and experimentation, technology can suddenly move forward at an alarming rate; within 15 years, robots and artificial intelligence will dominate the legal practice, and may bring 'structural collapse' to law firms, drastically changing the landscape of the legal services market." Susskind, a British scholar, who has researched technology and law for over 30 years, argues in his book *Tomorrow's Lawyer: an Introduction to Your Future* that the law industry will change more in the next 20 years than it has in the past 200 years. Legal professionals need to be prepared for the future. The questions of who bears obligations and who bears responsibility for harm caused are ones that the EU will need to focus on when it addresses robot legislation in the future.

This statement is not empty rhetoric. The legal profession is not completely immune to technology. In the face of technological development and external pressure, the legal profession has revealed its non-adaptability in many aspects such as educational model, organizational structure, and payment model. This gives people high hopes for LawTech with the support of artificial intelligence technology. Globally, although from 2011 to 2016, the total scale of financing for global LawTech companies was only USD 739 million, which is significantly lower than those in emerging fields such as FinTech and MedTech, the number of global listed LawTech companies has exploded, from 15 in 2009 to 1164 in 2016. The focus is on nine domains: online legal services; electronic forensics; business management software; intellectual property/trademark software services; artificial intelligence-enabled LawTech; litigation finance; legal search; legal advice, and notary tools. In the context of these international trends, China's domestic LawTech market has started to shift from "Internet + law" to "artificial intelligence + law," and entrepreneurship in the latter field has become an important part of artificial intelligence entrepreneurship. "AI + law" products for businesses and individual consumers are gradually entering the public purview.

Not only that, the legal profession, including law firms, corporate legal departments and courts, has also begun to aggressively deploy artificial intelligence LawTech, partly because of online legal services and customer cost pressures (e.g. corporate legal departments increasingly looking to get more legal services at a lower cost) as well as other factors that force law firms to invest in innovation. International law firm Dentons is the quintessential example. In May 2015, it took the lead in launching NextLaw Labs, a legal technology innovation accelerator project, which has incubated more than ten LawTech projects, including the famous robotic lawyer ROSS. More and more international law firms such as Linklaters, Riverview Law, BakerHostetler and others have also started to develop and deploy artificial intelligence systems to help them work more efficiently or provide legal services through a low-cost model.

In sum, from the earliest rule-based expert legal system (which incorporated the knowledge and experience of legal experts in the form of rules for converting to computer languages) to autonomous systems supported by deep learning, machine learning, big data and so on, artificial intelligence's deeper and broader impact on law and the legal profession has only just begun. It can be said that artificial intelligence technology is already beginning to transform the entire legal profession, and the scale and speed of the transformation will not only depend on the pace of technological development and progress, but also on the entire legal community's acceptance of new technologies and new models, which require policy support and development guidance. Based on some previous observations and studies, this author attempts to sum up the following ten major trends in the application and impact of artificial intelligence in the legal profession:

First, intelligentization and automation of legal discovery will profoundly affect the way legal practitioners conduct legal research (discovery).

With the aid of artificial intelligence technology, legal research (discovery) is moving towards intelligentization and automation. The value of legal research to legal practitioners is self-evident, whether you are a law school student, a practicing lawyer, a member of a corporate legal team, or part of the judiciary. Sometimes even an ordinary person also needs to search for something related to the law. In fact, informatization has already transformed legal search, and the digitization of legal documents such as legal texts and reference documents has buttressed the market for large legal databases. Still, legal database services such as Westlaw and *Beida Fabao* are generally based on traditional keyword search.[1] Using these databases for legal search is a time-consuming and laborious task. However, semantic retrieval and legal questions and answers based on natural language processing (NLP) and deep learning have begun to transform traditional legal retrieval services. For example, ROSS, which claims to be the world's first robotic lawyer, is an intelligent search tool based on IBM's Watson system. It uses powerful NLP and machine learning techniques to present lawyers with the most relevant and valuable legal answers, rather than the mass of search results given by traditional legal databases. In addition, semantic technology, text analysis and NLP, as well as image and video technologies, have opened up possibilities for the automation of intellectual property laws such as trademark and patent search and copyright monitoring, such as performed by the company TrademarkNow.

The new form of intelligent legal research based on voice interaction will go through two stages. The first stage is intelligentization. At this stage, there is still a need for human lawyers to clarify what legal issues need to be resolved or answered. The legal search engines identify relevant cases, and the lawyers assess their value to compose professional responses. ROSS is a typical representative of intelligent legal search at this stage. The second phase is automation, which means that no human lawyer is required

[1] A database providing services for those engaging in legal practice, study and research, jointly launched by Peking University Legal Information Center and Beijing ChinaLawInfo Co. Ltd.

to indicate what the legal problem is. The system itself can understand a factual statement and automatically identify the legal issues, then complete the search and provide the best legal information. The whole process requires almost no in-depth participation by human lawyers. This essentially frees human lawyers from cumbersome legal search work.

Second, artificial intelligence will continue to drive the automation of legal documents.

Just as the rise of news-writing robots will bring a huge change to the journalism industry, the trend of legal document automation will likely bring change to the legal industry that is on an equivalent scale or even more far-reaching. This will involve two main levels.

The first level is the automation of legal document review. Whether it is investigation and evidence collection, due diligence, contract analysis or compliance review, legal documents need to be reviewed, analyzed, and studied. For this work, automation will significantly improve the efficiency of legal professionals. Take the collection of electronic evidence as an example. Massive amounts of electronic materials pose great challenges for the collection and organization of evidence and legal materials in an increasing number of mergers and acquisitions, anti-monopoly cases, large-scale labor disputes, and other cases. Law firms often need to invest a lot of labor resources, material resources, and time. However, electronic forensics procedures based on technologies such as NLP, Technology Assisted Review (TAR), machine learning, and predictive coding can significantly improve the efficiency of this work, greatly reduce the time for reviewing documents, with accuracy no lower than human lawyers. Therefore, these technologies have become a large segment of the LawTech market, and companies such as Microsoft have been involved. The steps of electronic forensics generally include a training process (a human attorney confirms relevant evidence from a small sample for machine learning) and a forensic process (meaning that the machine replaces a human lawyer for data review to find evidence material). Because the use of machines to replace lawyers may come up against policy barriers, courts in countries such as the United Kingdom, the United States, and Australia have made it clear that predictive programming techniques can be used in gathering and collating evidence in litigation and cases.

Another major area of automation for legal document review is contract analysis. Contract analysis is of great significance in many situations, such as risk control, due diligence, forensics and litigation, but it is a time-consuming and labor-intensive job. Thus, Deloitte relies on the machine learning contract analysis system, Kira Systems. It takes only 15 minutes to read through a contract that would have taken 12 hours for a human lawyer to review. In the international community, artificial intelligence contract analysis services have been regularized, with an increasing number of LawTech companies providing intelligent contract services such as KMStandards, RAVN, Seal Software, Beagle, and LawGeex. Driven by artificial intelligence technology, artificial intelligence contract analysis services are still booming, bringing efficiency, cost reduction, and process improvement.

The second level is the automation of legal document generation. The journalism industry is being transformed by the Internet and machine writing. In the past eight years, journalism revenue has decreased by one-third, employment has decreased by 17,000, and the market value and power of newspapers have been greatly reduced. Instead, online media has continued to rise. The legal profession is facing the same situation, and the era of intelligent machine assistance and even independent drafting of legal documents will materialize.

Nowadays, the legal practitioner's formatting of legal documents is changing from the use of a template to taking place in an automated fashion; perhaps in the next 10–15 years, artificial intelligence systems will probably draft most transaction documents, legal documents, and even indictments, memos, and judgments. The role of the lawyer will change from a drafter to a reviewer. For example, a program developed by Fenwick & West, a law firm in Silicon Valley, can automatically generate the required documents for a startup that is ready to go public, which reduces the lawyer's billing time from 20–40 hours to a few hours. When a large number of documents need to be prepared, this program can reduce the time required from days and weeks to hours, greatly improving work efficiency. The advantage of machine intelligence is that as data is accumulated, it can continuously learn and improve itself, and because of the interrelatedness of data, computers can associate specific contracts with all relevant court decisions to form a dynamic relationship that continuously improves the legal format. In the future, with the continuous improvement of hardware and software performance and algorithms, high-level legal documents such as indictments, memos, and judgments can be automatically generated, but these will still require the review of human lawyers or judges, thereby forming a collaborative human-machine relationship.

Third, alternative business models such as online legal services and robotic legal services are emerging, making the provision of legal services increasingly standardized, commoditized, automated, and democratized.

In the Internet age and the era of artificial intelligence, law firms and human lawyers are not the only channels for the general public to obtain legal services. Alternative business models such as online legal services and robotic legal services are emerging to provide general legal advice to end users, such as for wills, marriage counseling, traffic accident advice, and more. DoNotPay, a legal robot for end consumers, can assist users to complete the preparation and submission of complaints about traffic tickets.

Richard Posner, an American jurist, once described the legal profession as a "cartel of providers of services related to society's laws," meaning a monopolistic industry. High legal fees have led to a large unmet legal need in society, and the legal needs of the majority of low-income and middle-income people have not been met. However, alternative business models such as online legal services and robotic legal services can provide legal services to end users at a lower price, which is expected to standardize, commercialize, automate, and democratize legal services. Commoditization means that the provision of legal services no longer depends primarily on the professionalism of specific human lawyers, but can be provided in an automated manner; democratization means that most people will be able to obtain general legal services at a lower cost. A British scholar, Richard Susskind, believes that the evolution of the provision of legal services from customization to standardization to systemization to comprehensive packages and finally to commoditization means that the pricing of legal services is decreasing, that is, from hourly billing to fixed fees to commercial pricing, ultimately trending to zero. At this level, some foreign experts predict that demand for lawyers will decline.

In any case, legal robots will have a profound impact on the provision of legal services and will continue to promote the standardization, systematization, commodification, and automation of legal services, so that everyone can obtain legal services and help eliminate the asymmetry of legal resources, realizing a broader notion of justice. Today, in the United States, the most well-known legal brand is not a well-known law firm but online legal service providers such as LegalZoom. These new technology-based legal service providers represent the future trend of legal service provision. They are not substitutes for the law firm, but rather outside the law firm, meeting other unmet legal needs or legal requirements that are expensive if handled by the law firm. The United Kingdom passed the Legal Services Act as early as 2007, aiming to liberalize the legal market, reform the legal industry organization model, and introduce competition to promote the affordability of legal services. In this context, some international law firms have established low-cost legal service centers to provide legal services at lower prices through technological assistance, in addition to hourly billing and fixed fees.

Fourth, case prediction based on artificial intelligence and big data will profoundly affect the litigation-related behavior of the parties and the resolution of legal disputes.

From the prediction of case results to crime prediction, predictive technologies based on artificial intelligence and big data are increasingly used in the judicial field. There has been much research progress in case prediction techniques. In 2016, researchers used the European Court of Human Rights' open judgment records to train an algorithm to predict the outcome of a given case, with a prediction accuracy of 79 percent. This empirical study shows that case facts are the most important predictor, and this conclusion is consistent with the view of legal formalism that judicial decisions are mainly influenced by the case's statements of facts. Case predictions have been used in many practical areas. For example, Lex Machina provides services that predict the outcome of a case by natural language processing of thousands of judgments. The software can determine which judges tend to support the plaintiff, form a litigation strategy based on past cases handled by the opposing lawyer, form the most effective legal argument for a particular court or judge, and so on. Lex Machina's technology has already been used in patent cases.

The value of case prediction is mainly reflected in two aspects. First, it can help the parties to form the best litigation strategy, thus saving litigation costs. Second, it can help judges to realize the "same-case, same-judgement" standard, that is, so-called big data judgements ensure fairness and justice. The potential high costs of litigation can impose a heavy financial burden on the parties, so the parties generally assess the likelihood of a successful case before litigation or before an appeal. However, even the most professional lawyers are far less predictive than computers because of the limited information processing capabilities of the human brain. Because computers are supported by powerful algorithms, they can handle almost all the data that can be obtained with superior computing power.

The computer's full data processing makes the case predictions more reliable than human sample data analysis. If you can predict the outcome of the case more reliably in advance, it means that the parties will not continue to promote litigation or appeals with a great risk of losing the case, but will choose other forms of resolution such as settlement and abandonment of litigation. However, the drawback of case prediction is that it may distort the litigation behavior of the parties and bring new prejudices and abuses.

Fifth, the online court, as well as artificial intelligence legal aid, will promote access to justice and help bridge the justice gap.

As the saying goes, "the door to the court opens to the south, if you are in the right but have no money, don't come in" (法院大门朝南开, 有理没钱别进来). The inefficiency, procedural delays, and high costs of the judicial trial system have traditionally been criticized. But the question is, do people have to go to physical court locations in order to resolve legal disputes? The development of technology has answered in the negative. For example, with the rise and prosperity of e-commerce, Online Dispute Resolution (ODR) became popular. On the e-commerce platform eBay, a large number of sales disputes were resolved online through SquareTrade, an ODR service provider. The parties submit factual statements and evidence online through the ODR system, so that disputes can be processed online without even involving human lawyers. Many cases will not get to a court trial at all.

Under the influence of the ODR model, the practice of online courts has emerged abroad. For example, Lord Briggs, a judge of the Court of Appeal of England and Wales, said in his call for "improving the efficiency of civil justice" that artificial intelligence can help with online rulings on civil law cases in England and Wales. In this respect, artificial intelligence can assist judges even in passing judgements. It is reported that the United Kingdom has invested GBP 1 billion to modernize and digitize its court system. According to the British scholar Susskind, the online court in the United Kingdom consists of three phases: the first phase is the online legal aid system, which provides legal advice and advice to the parties; the second phase is pre-trial dispute resolution, where the judge communicates with the client by email or telephone to resolve the dispute; the third phase, online courts, only applicable to small cases, conduct online trials, including filing, submitting evidence, evaluating evidence, and making a decision. This is similar to a summary proceeding. The current online court in the United Kingdom does not use artificial intelligence systems to decide cases, so it is not a substitute for judges, but helps provide a better way to resolve disputes. In the context of increasingly digitized communication scenarios, online identification, audio and video technology, and artificial intelligence technology have provided technical support for the construction of online courts.

The smart courts that China is vigorously promoting are similar to the online courts abroad. Per the "Outline of the National Informatization Development Strategy" released in July 2016, the construction of "smart courts" will be included in the national informatization development strategy, and it is clearly stated: "Build 'smart courts,' raise the informatization levels of all stages, including acceptance of a case, trials, judgments, enforcement and supervision, promote the openness of judicial and law enforcement information, and push forward judicial fairness and justice." The "13th Five-Year National Informatization Plan," issued in December 2016, clearly pointed out that it supports the construction of "smart courts," promotes electronic litigation, and builds and improves upon a fair judicial informatization project. At the Fourth Informatization Work Conference of the National Court held on May 11, 2017, Zhou Qiang, President of the Supreme People's Court, proposed that smart courts are a form of the court's work based on the foundation of informatization. Courts across China are exploring some form of smart court construction, but the best-known case is the construction of the Zhejiang smart court (Zhejiang E-Commerce Court). According to Liu Keqin, deputy director of the Zhejiang High Court's information center, the Zhejiang smart court handles up to 23,000 disputes in transactions and copyrights each year. It can directly dock multiple platforms such as Taobao and Tmall, and provide a platform for diversified online conflict resolution. Other auxiliary measures include prejudging case results, conducting online judicial auctions, intelligent speech recognition, matching cases to similar cases, pulling up credit profiles of party members, and so on. On June 26, 2017, the Central Comprehensively Deepening Reforms Commission reviewed and approved the "Proposal on Establishing Hangzhou Internet Court", which mainly deals with five types of cases: online shopping contract disputes, online shopping product liability disputes, network service contract disputes, financial loan contract disputes, and microfinance contract disputes.[2] In the future, the further construction and popularization of online courts will promote the supply of public legal services and help eliminate the judicial divide.

In addition, the lack of legal aid in public legal services is also a major problem in the judicial system. Especially in criminal cases, many defendants do not receive legal consultation and defense. Some civil cases were also carried out without the intervention of a lawyer. In the future, legal robots can provide basic legal aid to the parties—with legal aid lawyers only intervening when necessary—which can significantly improve the efficiency and quality of judicial assistance and achieve fairness and justice. Moreover, the provision of legal aid through legal robots can also be integrated into the construction of online courts.

Sixth, artificial intelligence and robots will become the main point of entry to the legal system.

Whether it is a law firm or a lawyer, a court, or a client or end consumer, the "intelligence interface" based on artificial intelligence and robotics will become the main entry point of the legal system. Legal robots and artificial intelligence will be at the core. For lawyers, future legal practices such as legal search, case management and legal writing will be mainly done through legal robots and artificial intelligence systems with intelligent interactive interfaces, just as doctors now rely on a variety of complex devices to complete medical activities. For the court, the digitization of judicial trials means that case retrieval, judgment writing, evidence analysis, and reasoning will also be carried out or even replaced by legal artificial intelligence. For end users, an interactive, Internet-based question-and-answering system can communicate with users in the form of text or voice conversations and generate the required legal information or guide them through the completion of basic legal documents and formats. In this context, the current role of lawyers will change, and machines may replace some roles, such as routine, repetitive tasks; machines may also enhance some roles, such as case prediction and legal writing. For new laws and regulations, lawyers still need to play a central role.

[2] The Central Comprehensively Deepening Reforms Commission was formerly known as the Central Leading Group for Comprehensively Deepening Reforms, which was formed in December 2013 with China's President Xi Jinping as its leader. The group is tasked with reforming the economic, political, social, and party-building systems.

Seventh, evaluations of lawyers will make the legal market more transparent and may bring about the "Matthew Effect."

As a two-sided market, the legal market's evaluation system is largely opaque, unlike those for e-commerce platforms and the O2O platform such as food delivery services, which have relatively complete user evaluation mechanisms to ensure market transparency and the consumer's right to knowledge. Since the legal market is largely not platform-ized, it is difficult to build an effective evaluation mechanism. However, artificial intelligence and big data are changing this situation, and it is becoming possible to better evaluate the market for lawyers—a major trend in legal science and technology. At present, recommendations for lawyers has become one of the core areas of LawTech, and products and services for recommending and evaluating lawyers have continued to emerge at home and abroad. The evaluation of the market for lawyers is equivalent to placing the lawyer in the sun. The distinction between star lawyers, ordinary lawyers, and unqualified lawyers will become clear. This may bring about a "Matthew Effect" in the market for lawyers. The business and income of star lawyers will increase, while the opposite will occur for the average and junior lawyers. This calls for a transformation of the work of lawyers, that is, towards providing legal services in a technologically enabled, low-cost model.

Eighth, the legal artificial intelligence profession will come to the fore as an emerging profession within the legal industry.

Legal robots and legal artificial intelligence are not created out of thin air and require the cooperation of technicians and legal experts. With the continuous integration of artificial intelligence and law, the research, development, and application in this field will continue to increase, and the legal artificial intelligence profession will come to the fore as an emerging profession within the legal industry. At present, some international law firms that actively embrace new technologies are already building up their legal IT capacity. Code developers for legal applications, legal data analysts, and legal database managers are joining law firms, corporate legal departments, courts, legal database companies, and other legal institutions. Legal technology companies need a combination of people who understand both law and technology. In the future, technology and law will be more closely integrated, and the demand for new talents will be more urgent.

Ninth, legal education and cutting-edge information science and technology, such as artificial intelligence, will be increasingly tied together.

China's "New Generation Artificial Intelligence Development Plan" noted the integration of legal education and artificial intelligence and proposed to create a new composite model of "artificial intelligence + law" professional training. This is an extremely forward-looking vision. This author has participated in the translation of the book *Failing Law School*, which sharply criticizes the legal education model of "4+3" (four-year undergraduate + three-year law school education) in the United States and believes students do not need to attend three years of law school. Two years or maybe one year is enough. On the other hand, traditional Chinese legal education consists of directly studying the law for four years in undergraduate school after graduating from high school. Such a training model for legal talents will meet difficulties in adapting to the future legal practices guided by robots and artificial intelligence. Compared to current lawyers, future lawyers will work in very different jobs and therefore require different models of education. Therefore, the "artificial intelligence + law" training model proposed by the new plan is far-sighted.

In fact, international law schools have long begun to explore innovative legal education, focusing on the cultivation of science and digital literacy for law students. For example, as early as 2012, the Georgetown University School of Law began offering a practical course in technological innovation and legal practice, forming a distinctive "Iron Tech Lawyer" competition to foster students' abilities in developing legal apps. In 2015, the University of Melbourne Law School began offering courses on how to develop legal apps. In the future, legal education and cutting-edge information science and technology, such as artificial intelligence, will be increasingly tied together, and whether this idea can be realized earlier and faster depends on the legal education system's response speed. In fact, artificial intelligence not only challenges legal education but also necessitates an interdisciplinary educational model, posing similar challenges to other disciplines.

Tenth, computational law and algorithm judges will become the ultimate form of law.

In the online court proposal, Briggs of the Court of Appeal of England and Wales, put forward the notion of an algorithm judge, that is, artificial intelligence can directly make a judgement instead of a judge. This is not impossible. In fact, computational law has always been one of the core research directions of the artificial intelligence and law field. Consider the question: "Setting aside written language, are there more precise and more formal ways to express the law?" This leads to the exploration of computational logic and code to express the law. I have previously seen an idea on Zhihu: If one can use a list of n-dimensional vectors to describe various events, then one can import the "event.txt" into "legal.exe", resulting in "decision.txt".[3] The legal provisions are translated into code, so that the judgment is completely insulated from personal subjective judgments and can also be conducted online on anyone's computer. Then, one would make the code open source and put it on a GitHub-like website for public supervision. Computational law is currently used in some areas such as taxation, but it is more discussed as an academic research direction. However, in the mature information society of the future, more widespread computational laws will emerge, and the system will automatically enforce the law without a lawyer. Even the judge will not be needed, because by that time, the law will be fully automated.

Legal Practitioners Should Be Prepared for the Future

People say that the best way to foresee the future is to create the future. The future of the legal profession must be created by the legal community acting together. Previous studies have found that the probability of paralegal and legal assistants being automated is as high as 94 percent, raising concerns about the employment of law graduates. However, the author's test results on willrobotstakemyjob.com show that only 3.5 percent of lawyers will be replaced by artificial intelligence and robots. Whether this figure is scientific or not, it can provide some temporary relief.

According to this author's investigation, a lawyer's work includes 13 items: document management, case management, document review, due diligence, document drafting, legal writing, legal search (and research), legal analysis and strategy, fact investigation, customer advisory services, other exchanges and interactions, appearances in court, and preparations. Lawyers need to think early about which of these tasks can be automated or can become more efficient through technology. The British scholar

[3] Zhihu is a Chinese site similar to Quora.

Susskind puts forward the idea of "decomposing" legal services, thinking that a legal task can be broken down into multiple parts. The core part can be done by a lawyer and the rest by a more efficient third party.

As for those concerned about the automation of legal services, including lawyers and other legal practitioners, at least three factors need to be considered in judging the value of legal work and thinking about the impact of artificial intelligence technology on this work. First, whether the work involves data analysis and processing. If so, it is almost impossible for humans to compete with artificial intelligence and robots, and it is sensible to use and adapt to new technologies as soon as possible. Second, whether it involves interactive communication, similar to the legal front desk and other legal customer service work. If so, the probability of automation is extremely high. Standard legal advice can also be automated, but higher levels of interactive communication such as negotiation and court appearance are difficult to be automated in the short term. Third, whether it supports decision-making. Artificial intelligence–assisted decision has been applied in many fields. In the legal industry, artificial intelligence–assisted decision-making is also occurring and becoming a trend. For example, in the prediction of case results, it is possible for artificial intelligence to perform better than professional lawyers. In this way, it is an inevitable choice to use and adapt to new technologies as early as possible.

Finally, as a summary, after more than 30 years of development, influenced by superior computing power, big data, and continuous improvement algorithms, the impact of artificial intelligence on both the law and the legal industry is deepening and accelerating. In the next 10–20 years, the legal industry will likely undergo a major change. As the most direct target customer of legal artificial intelligence, legal practitioners need to adjust their mentality, actively embrace new technologies and new models, and uphold legal concepts and beliefs in this process. This is needed to prevent legal artificial intelligence from weakening and damaging the concepts and values that the legal community and system uphold, and to let legal artificial intelligence promote justice, rather than lead to prejudice and discrimination, or run counter to and degrade justice.

PART V

Ethics: Human Values and Human-Machine Relations

With the third wave kickstarting the era of artificial intelligence, artificial intelligence ethics has once again become one of the core topics of hot discussion and research. The United Nations, the European Union, the United States, and the United Kingdom have paid special attention to this issue and are introducing various measures such as research reports, guidelines, and laws and policies to promote awareness and resolution of ethical issues in artificial intelligence. The Institute of Electrical and Electronics Engineers (IEEE), the Internet Society, the Asilomar Conference, and Google, IBM, and other organizations have also begun to actively respond to artificial intelligence ethics in the form of ethical standards, artificial intelligence principles, artificial intelligence ethics review committees, and other formats of industry self-regulation to actively face the issue of artificial intelligence ethics.

It can be foreseen that with the rise of machine intelligence, artificial intelligence, robots, and other technologies, engagement will involve more and more ethical decision-making. In the future, strong artificial intelligence and super artificial intelligence will bring about the problem of human-machine differentiation, and strengthening the research of artificial intelligence ethics will become more important, particularly to preserve the value of human beings and to achieve the vision of coexistence and co-prosperity.

CHAPTER 23

Moral Machines

"We had better be quite sure that the purpose put into the machine is the purpose which we really desire."
—Norbert Wiener[1]

2016 was an exceptionally bright year in the history of artificial intelligence. This year, Google's DeepMind team developed the Go-playing program AlphaGo that defeated top human players for the first time, which would not have been possible without the contributions of AI technology such as deep learning and reinforcement learning. The following year, AlphaGo swept almost every top human player on multiple occasions, and the era of human rule in Go completely ended. Another robot program, Libratus, beat top human players in a Texas Hold'em game. This was the first time that robots have defeated humans in a game of incomplete information. These events mark the rise of machine intelligence, and human society is gradually entering the era of intelligent machines. It will become increasingly common for machines to assist or even replace humans for various decisions.

The symbol of this artificial intelligence wave is deep learning. It is a self-learning, self-programming learning algorithm that can be used to solve more complex cognitive tasks. These tasks have previously belonged

[1] Norbert Wiener was a renowned mathematician who wrote on the AI control problem in 1960.

Tencent Research Institute et al. (eds.), *Artificial Intelligence*,
https://doi.org/10.1007/978-981-15-6548-9_23

exclusively to human or human experts, such as driving, identifying faces, providing legal consulting services, and more. Deep learning, reinforcement learning, and other machine learning technologies, combined with big data, cloud computing, Internet of Things (IoT), and other hardware and software technologies, have made major breakthroughs in machine intelligence. In this context, the calls for moral machines have been renewed.

The Accelerated Arrival of Intelligent Machines

Artificial intelligence technology helps accelerate the arrival of intelligent machines, as machines are gradually changing from passive tools to active ones. "Computers can only execute mandatory instructions—they are not programmed to make it possible for them to make judgments," said a court in New York. This may represent the public's view of computers and robots. However, advances in artificial intelligence technology are making this argument outdated, and it may even become a biased preconception. Because machines are changing from passive tools to active ones, they can possess the abilities to perceive, recognize, plan, make decisions, and carry out decisions as humans do.

Since 2010, a number of mutually reinforcing factors, including big data, continuously improving machine learning, more powerful computers, and the informatization of the physical environment (through IoT) have driven the rapid development of artificial intelligence technologies in the Information and Communication Technology field. These are being applied to more and more fields and scenarios such as self-driving cars, medical robots, nursing robots, industrial and service robots, and Internet services. Some foreign insurance and financial companies and law firms have even begun to replace human employees with artificial intelligence systems possessing cognitive capabilities. From chess and quizzes (such as Jeopardy), to Go and Texas Hold'em, to medical diagnosis, image, and speech recognition, artificial intelligence systems have begun to reach or even exceed the cognitive level of human beings in more and more fields. Allowing them to assist or even make decisions in the place of humans is no longer a utopian fantasy. It is now reasonable to foresee that in the near future, various intelligent machines or smart robots in various fields such as transportation, medical care, nursing care, industry, and service industries will become common features of human society.

In the case of a self-driving car, the greatest feature distinguishing it from traditional machines is its high degree of, or even complete, autonomy. No matter what machine learning method is used, the current mainstream deep learning algorithm does not program the computer step by step, but allows the computer to learn from the data (usually a large amount of data) without requiring the programmer to make new step-by-step instructions. Therefore, in machine learning, learning algorithms create rules rather than programmers. The basic process is to provide training data to the learning algorithm. Then, the learning algorithm generates a set of new rules, or a machine learning model, based on the inference obtained from the data. This means that computers can be used for complex cognitive tasks that cannot be programmed manually, such as image recognition, translating pictures into speech, driving cars, and so on. In the case of self-driving cars, they use a series of radar and laser sensors, cameras, global positioning devices, and many sophisticated analytical procedures and algorithms to drive cars like humans and even do better than them. Self-driving cars "observe" the road conditions, continue to pay attention to other cars, pedestrians, obstacles, bypasses, and so on, take into account traffic flow, weather, and all other factors that affect the safety of driving a car, and constantly adjust the speed and route. In addition, self-driving cars are programmed to avoid collisions with pedestrians, other vehicles or obstacles. All of this is the result of machine learning. Therefore, it can be said that in every realistic situation, the self-driving cars themselves are independently judging and making decisions, although the programmers set the learning rules.

Further, intelligent machines may "break" pre-set rules that greatly exceed their designers' expectations. People have always been concerned that giving the machine the ability to "think" autonomously may lead to its ability to violate established "rules" and behave in unexpected ways. This is not purely imagination. There is already evidence that highly "intelligent" autonomous machines can learn to "break" rules to protect their own survival. The learning and experience of a self-driving car after it leaves the manufacturer's control and enters circulation also influence its behavior and decision-making. The new data inputs may make the self-driving car adjust and adapt, causing its behavior and decisions to overstep the pre-set rules. This is not theoretically impossible.

These phenomena all indicate that computers, robots, machines, and so on are moving away from direct human control and operating autonomously, although they still require human beings to start them up, and are indirectly controlled by humans. In essence, self-driving cars, intelligent robots, and various kinds of virtual agent software are no longer passive tools in human hands, but agents of humanity, with autonomy and initiative. This presents a major challenge to ethics and morality. Previous ethical norms for humans and human society now need to be extended to intelligent machines, and this may require a new ethical paradigm.

The Need for Moral Code

Future autonomous smart machines will have the ability to act completely autonomously and will no longer be passive tools for humans. Although humans design, manufacture, and deploy them, their behavior will not be constrained by direct human instructions but based on analysis of and judgments about the information they acquire. Moreover, their reaction and decision-making in different situations may not be predictable or controlled in advance by their creators. Full autonomy means a new machine paradigm—"perception-thinking-action"—that does not require human intervention or any other intervention per se. This shift means new ethical requirements for artificial intelligence, robots, and so on, and calls for a new ethical paradigm for machines.

When the decision-maker is human and the machine is only the tool of human decision-makers, humans need to be responsible for their use of the machine's actions. They have the legal and ethical obligations of good faith, reasonableness and proper use of the machine. Morally, they must not use the machine as a tool to engage in wrongdoing. In addition, when human decision-makers resort to tools to engage in improper or illegal acts, human society can condemn on the grounds of morality and public opinion. Alternatively, they can use the law itself as a tool to punish lawbreakers. However, existing legal and ethical paths for human decision-makers do not apply to non-human intelligent machines. Since smart machines themselves, prior to replacing humans, can only act according to decisions made by humans, when designing smart machines, people need to put forward similar legal and ethical requirements for smart machines as dynamic agents. This is needed to ensure that the decisions made by smart machines can be just like those made by humans, and be ethical and legal, as well as having corresponding external constraints and sanctioning mechanisms.

Further, some issues in intelligent machine decision-making also highlight the importance of machine ethics. It is necessary to make a highly autonomous intelligent machine a moral entity like human beings, that is, a moral machine. One of the problems is that since a deep learning algorithm is a "black box", how the artificial intelligence system decides is often not known, and it may hide many problems such as discrimination, prejudice, and inequality. The increasingly prominent issues of discrimination and injustice in artificial intelligence decisions make AI ethics especially significant. In particular, artificial intelligence decisions have been widely used in areas such as driving, loans, insurance, employment, criminal investigation, judicial trials, face recognition, and finance. These decision-making activities affect the vital interests of users and people. It is crucial to ensure that smart machine decisions are fair, reasonable, and legal, because maintaining everyone's freedom, dignity, safety, and rights is the ultimate pursuit of human society.

In addition, artificial intelligence in warfare should be particularly subject to ethical norms. At present, many countries are actively researching and developing military robots, and an important development trend of military robots is the continuous improvement of autonomy. For example, the X-47B Unmanned Aerial Vehicle developed by the US Navy can achieve autonomous flight and landing. Countries such as South Korea and Israel have developed sentry robots. They have automatic modes and can decide whether to fire or not. Obviously, if military robots are not controlled in a certain way, they are likely to have no sympathy for human beings and will be merciless to achieve their goals. Once they are started, they may become truly cold-blooded "killing machines." In order to reduce the harm that military autonomous robots may cause, they need to be made to abide by human ethical standards, such as not harming noncombatants and distinguishing between military and civilian facilities. Although there are still some difficulties in the current technology to achieve such goals, technical difficulties do not mean negating their necessity and possibility.

Realizing Moral Machines

Artificial intelligence systems such as robots and intelligent machines need to comply with and be bound by human society's norms including moral and legal ones. However, how to achieve this goal, which is to design moral machines that embed the norms and values of human society's laws and ethics into artificial intelligence systems, is a big challenge. First of all, people need to ask, can legal and moral requirements and norms be converted into computer code, that is, is moral and ethical computer code possible? Second, if so, what are the specifications and values that need to be embedded? And how should these legal and ethical requirements be embedded in artificial intelligence systems? Finally, how can we ensure that the norms and values embedded in artificial intelligence systems are in line with human interests and keep up with the times? Solving these three problems can basically ensure the realization of machine ethics and make artificial intelligence systems into agents whose behavior is as well-meaning, honest, and lawful as that of humans.

In order to solve the problem of ethical embedding, at the end of 2016, the Institute of Electrical and Electronics Engineers (IEEE) initiated its artificial intelligence ethical project and released "Ethically Aligned Design: A Vision for Prioritizing Human Well-being with Autonomous and Intelligent Systems (AI/AS)" from the perspective of operational standards. It provides guidance for ethical embedding and is worth exploring and learning from. The IEEE divides the implementation of artificial intelligence ethics into three steps.

First, identify the norms and values of specific communities. To start with, the norms and values to be embedded in the AI system should be clarified. Legal norms are generally written, formalized, and easily identifiable. However, social and moral norms are more difficult to confirm. They are reflected in behaviors, languages, customs, cultural symbols, handicrafts, and so on. Further, norms and values are not universal, and the values that need to be embedded in AI should be a set of norms for specific tasks in a particular society or group.

Second, there is the issue of moral overload. AI systems are generally subject to a variety of norms and value constraints, such as legal requirements, monetary benefits, social and moral values, and they may conflict with each other. What values should be placed in the highest priority in these situations? Therefore, priority should be given to the value system shared by the majority stakeholder groups; when determining the ordering of values in the AI R&D phase, there need to be clear and explicit justifications; the ordering of values may change over time and the technology should reflect this change.

Finally, there are data or algorithmic discrimination issues. The AI system may intentionally or unintentionally cause discrimination against specific users. On the one hand, it must be acknowledged that AI systems are prone to internal discrimination. We should be aware of the potential sources of this discrimination, and adopt more inclusive design principles. This should be strongly encouraged throughout the engineering stage, from design to execution to testing to marketing, being as inclusive as possible and including all prospective stakeholders. On the other hand, to maintain transparency in the resolution of value conflicts, it is particularly necessary to consider the interests of disadvantaged and easily overlooked groups (children, the elderly, offenders, ethnic minorities, poor people, people with disabilities, etc.). In the design process, we should take an interdisciplinary approach and involve relevant experts or consultants.

The second step is embedding the norms and values identified into artificial intelligence systems. After the specification system is confirmed, how to build it into the structure of the computer is a problem. Although related research has been continuing in fields such as Machine Morality, Machine Ethics, Moral Machine, Value Alignment, Artificial Morality, safe AI, friendly AI, the development of computer systems that can recognize and understand human norms and values and allow them to consider these issues when making decisions has continued to trouble people.

There are currently two main paths: the top-down path and the bottom-up path. Wallach and Allen described the "top-down" approach as one that "takes the antecedently specified ethical theory and analyzes its computational requirements to guide the design of algorithms and subsystems capable of implementing that theory." The "bottom-up" approach is called a "developmental approach," which states that the approach "is focused on creating an environment for the subject to explore action and learning, and encourage them to implement ethically commendable behavior." They claim that the bottom-up approach has the advantage of being able to "dynamically integrate input from different social mechanisms" and can provide skills and standards for improving its overall development, but this method may have the drawback of being difficult to adapt and develop. It is not yet clear how these specifications will be embedded in computer architectures. Research in this area needs to be strengthened.

However, there may still be an ethical dilemma on which it is difficult to reach consensus, but which needs to be resolved in advance. Taking self-driving cars as an example, we can posit an ethical dilemma similar to the "trolley problem." Suppose a self-driving car fails to brake or brakes too late, and there just happen to be five people jaywalking in front of the car, and two passengers in the car. At this time, if the car continues to move forward, it will smash five people who do not obey the traffic rules, and if the car turns, it will encounter a roadblock, resulting in the deaths of the two people in the car. In this situation, how should people expect the car to choose? Since human ethical values are sometimes specious or conflicting, self-driving cars may find it difficult to make a proper choice at this time.

For example, according to utilitarianism, in order to maximize the interests and welfare of the greatest number of people, the car should sacrifice the two people in the car and save the five jaywalkers. However, in accordance with the moral requirements of absolutism, the act of harming a person in violation of one's free will is not permitted, and a person cannot be harmed against her free will in order to save the majority. In this context, this would cause the passengers in the car to lose their lives. Solving such problems is very important for the development and commercial application of artificial intelligence systems such as self-driving cars, so countries around the world are actively paying attention and responding.

The third step is assessing whether the norms and values embedded in artificial intelligence systems are consistent with those of humans. The specifications and values embedded in the AI system need to be evaluated to determine whether they are consistent with the real-world normative system, which requires evaluation criteria. Evaluation criteria include compatibility of machine specifications and human specifications, AI passing an approval process, trust in AI, and so on.

Establishing trust between humans and AI involves two levels. As far as users are concerned, the transparency and verifiability of AI systems are necessary to build trust; of course, trust is a dynamic variable in human-machine interaction and may change over time. When it comes to third-party evaluations, in order to promote the assessment of the system as a whole by third parties such as regulators and investigators, designers and developers should first record changes made to the system on a daily basis. This highly traceable system should have a model similar to the black box on the aircraft, which records and helps diagnose all the system's changes and behaviors. Second, regulators, together with users, developers, and designers, can define the minimum standards of value consistency and compliance, as well as the criteria for assessing AI reliability.

One of the more important issues in the ethical evaluation of artificial intelligence is actually value matching. Nowadays, many robots are single-purpose. Sweeping robots will concentrate solely on sweeping the ground. Service robots will wholeheartedly make you coffee, and so on. But is the robot's behavior really what we humans want? This creates a problem of value matching. The myth of King Midas provides an example. He wanted a technique that turned stone into gold. As a result, when he had this magic power, everything he touched, including food, would turn into gold. In the end, he starved to death. Why? Because this magic power did not understand the true intentions of King Midas. So will robots bring us similar situations? This issue is worth pondering.

Another potential scenario is that of your household robot killing your dog in order to cook for your child. More extreme, a robot that eliminates human suffering may find that humans may find ways to make themselves suffer even in a very happy environment. In the end, the robot may reasonably believe that the way to eliminate human suffering is to eliminate humans. This hypothesis has a realistic impact on medical robots, elderly care robots, and so on. Therefore, it was proposed that human-compatible AI covers three principles. The first is altruism, that is, the only goal of robots is to maximize the realization of human values. The second is uncertainty, that is, robots are at first not sure what human values are. Third is consideration of humans, that is, human behavior provides information about human values, thereby helping robots determine what values humans want. To solve the problem of value matching requires more interdisciplinary dialogue and exchange mechanisms.

Realizing Ethical and Moral Artificial Intelligence Requires a Comprehensive Governance Model

As mentioned earlier, demonstrating a moral machine mainly requires two aspects. One is the operational standards related to morality and ethics; the other is the methodology of ethical engineering. It is precisely because of the existence of these two problems that artificial intelligence ethics requires an interdisciplinary approach and method. Relying solely on humanities scholars or technical personnel will not suffice. This is because cross-discipline participation, dialogue, and communication will be absolutely necessary when dealing with ethical issues relating to artificial intelligence in the future.

In addition, just as humans acquire norms and values such as morality, law, and ethics through learning and social interactions, and apply self-discipline based on these norms and values, machine ethics hopes to achieve the same results. Through the establishment, implementation, testing, and inspection of ethical standards, we hope that the autonomous decision-making behavior of smart machines will respect the various norms and values of human society in advance and maximize the interests of humanity as a whole.

Considering human behavior, it is far from enough to only have human morality and legal self-regulation. An external monitoring and sanctioning mechanism is also required. Therefore, embedding ethics in artificial intelligence systems as a type of self-disciplining behavior is far from sufficient, and we need the participation of government regulators and the public to monitor, review, and provide feedback on the behavior of the artificial intelligence system in the course of an event or after the event, in order to both realize artificial intelligence ethics and ensure social fairness and justice. Therefore, the realization of artificial intelligence ethics is an all-round governance project. It requires AI R&D personnel, companies, governments, all sectors of society, and users to all play their respective functions and roles to ensure that artificial intelligence systems operate in a manner that respects and maintains the norms and values of the existing ethical and legal aspects of human society. While maximizing the benefits of artificial intelligence, they can also safeguard the freedom and dignity of the entire society and each individual.

Bibliography

Du, Yanyong. "现代军用机器人的伦理困境". *伦理学研究*. 5 (2014):98–99.

Wang, Donghao. "工智能体引发的道德冲突和困境初探". *伦理学研究*. 2 (2014a):70.

Wang, Shaoyuan. "论瓦拉赫与艾伦的AMAs的伦理设计思想:兼评《机器伦理:教导机器人区分善恶》". 洛阳师范学院学报. 33-1 (2014b):32.

CHAPTER 24

23 "Strong Regulations" for AI

Alan Turing, the father of computer science and artificial intelligence, once said: "Even if we could keep the machines in a subservient position, for instance by turning off the power at strategic moments, we should, as a species, feel greatly humbled." Turing wrote this passage in 1951, when the notion of artificial intelligence was not yet born. However, in today's age of rapidly advancing artificial intelligence technology and applications, academic leaders still have significant concerns about machines that are more intelligent than humans and may threaten humankind. Accompanying the fears of the possibility of artificial intelligence destroying humankind, government, industries, and enterprises are beginning to explore artificial intelligence "strong regulations"[1] and "controlling spells,"[2] aiming to make artificial intelligence beneficial to human beings as well as being safe, reliable, and controllable, so that it will not threaten the survival of our species.

[1] The Chinese phrase is军规 [jun2 gui1] which is a shorthand for军事法规 (jun1 shi4 fa3 gui1). My translation is "strong regulations," but alternative translations could include "military regulations" or "army rules." On the web, this Chinese phrase often appears as a reference to the Chinese title for the Joseph Heller novel *Catch-22.* This chapter's author uses this phrase not to refer to regulations on military applications of AI nor to refer to the actions by the military to regulate AI; rather it is used to connote strict rules to follow for the general development of AI. Thus, I translate it as "strong regulations" rather than the alternatives.

[2] The Chinese phrase "紧箍咒" [jin3 gu1 zhou4] could also be translated as "band-tightening spell," which is a reference to a magic spell used by the Monk in the novel *Journey to the West* to keep the Monkey King under control.

Tencent Research Institute et al. (eds.), *Artificial Intelligence*,
https://doi.org/10.1007/978-981-15-6548-9_24

Concerns Over Losing Control of Machines Are Long-Standing

In the summer of 1956, top computer scientists held a meeting in Dartmouth, in the eastern United States, and for the first time proposed the concept of "artificial intelligence." For the first time, at this meeting, it was decided that machines who think like humans should be referred to as "artificial intelligence." Since then, artificial intelligence has been a recurring theme of discussion, experiencing several high points and low points. Now that the third wave of artificial intelligence has arrived, the pace of its development will be significantly accelerated. In this process, whether it is the horrifying narratives of science fiction stories such as "The Terminator" and "The Matrix," the warnings of Hawking and other leading scientists, or the concerns of industry leaders such as Musk, all reveal humanity's fears of future artificial general intelligence and artificial superintelligence.

One can envision that artificial intelligence is moving from weak artificial intelligence to artificial general intelligence and artificial superintelligence. As long as technology continues to evolve, human beings will one day create artificial general intelligence, entering the phase of "intelligence explosion" or "technological singularity" proposed by mathematician I.J. Good. At that time, artificial general intelligence will have the ability to recursively self-improve, leading to the emergence of artificial superintelligence, whose upper limit is unknown. Kurzweil, who was hailed by Bill Gates as "the best person I know at predicting the future of artificial intelligence," predicts that robot intelligence will rival human beings by 2019; in 2030, humans will be combined with artificial intelligence into a "hybrid," and computers, connected to the cloud, will enter the body and brain. These cloud computers will enhance our current levels of intelligence. By 2045, human and machine will be deeply integrated, and artificial intelligence will surpass human beings and open up a new era of civilization.

Once humans are unable to effectively control future artificial general intelligence and artificial superintelligence, they may become the greatest threat to the overall survival and safety of humankind. Compared with nuclear technologies such as the atomic bomb, these types of threats will be worse than ever, so they require human prevention in advance. Although, as the White House AI report "Preparing for the Future of Artificial Intelligence" states, the current stage is the weak AI phase, and universal AI will not be available for decades to come; many researchers in the AI domain believe that as long as the technology continues to develop, artificial general intelligence and the subsequent artificial superintelligence will inevitably appear, with the main issue of disagreement being exactly when that general purpose artificial intelligence and artificial superintelligence will show up. Since 2016, celebrities such as Hawking, Elon Musk, and Eric Schmidt have all expressed worries about the development of artificial intelligence. They even think that the development of artificial intelligence will open the door to human destruction.

Hawking said in a speech, "I believe there is no deep difference between what can be achieved by a biological brain and what can be achieved by a computer. It therefore follows that computers can, in theory, emulate human intelligence—and exceed it." Elon Musk warned that if artificial intelligence is not properly developed, it may be like "summoning the demon." People are concerned that with the development of artificial general intelligence, humankind will usher in a "intelligence explosion" or "singularity"; at that time, the wisdom of machines will rise to an unforeseen level. When the wisdom of machines goes beyond human and superintelligent machines appear, humans may not be able to understand and control their own creations. The machines may revolt, which would be fatal and catastrophic to humans. This is a problem worth pondering. Of course, it is also necessary to undertake research ahead of time and take precautionary measures to ensure that artificial intelligence develops in a beneficial, safe, and controllable direction. This requires the formulation of "strong regulations" for artificial intelligence that give artificial intelligence a "controlling spell" to maximize the interests of humankind.

Are Asimov's Three Laws of Robotics Reliable?

No one was an earlier proponent of the issues of AI safety and ethics than Asimov. In many science fiction novels, Asimov often referred to safety and ethical standards for engineering robots. In the science fiction story *Runaround*, which came out in 1942, he proposed the three laws of robotics, in order to regulate robots ethically.

The first law is that robots must not harm humans, and must not sit idly by and watch human beings suffer harm. The second law is that robot should follow human instructions, unless the instructions are contrary to the first law. The third law is that without violating the first and second laws, robots should protect themselves. Later, Asimov revised the three laws of robotics and added a 0th law, which stated that the robot must protect the overall interests of humankind from harm. In evaluating the 0th law, people have said, "The messy idea of the overall interests of humankind cannot even be understood by humans themselves, let alone those robots who think in terms of 0s and 1s."

However, people always question whether the three laws of robotics can really solve the problems of robot safety and ethics. As many of Asimov's novels show, the flaws, vulnerabilities, and ambiguities of the three laws of robots will inevitably lead to some abnormal robot behavior. For example, in the movie "I, Robot," the VIKI robot ultimately decides to limit humanity's freedom in order to prolong the human species and stop wars among humans. However, according to the content of the three laws, this does not violate the three laws because the three laws do not define human rights, but merely ensure the safety of human life. Therefore, the action of the robots in the film to limit humans' freedom in order to protect them completely abides by the laws.

There are many robots in Asimov's Robot series of science fiction novels that illustrate the contradictions and conflicts among the three laws. One can see the defects and deficiencies of the three laws of robotics in building robot safety and ethics. The contradictions of the three laws of robotics can also be found in daily life. For example, when the police and gangsters are in a gunfight, according to the first law, robots cannot stand by and watch human beings suffer harm, so the robots must help both parties to ensure that they are not harmed. But is this the situation that humankind wants to see? Through many scenes such as these, we will see the shortcomings of the three laws of robotics.

Artificial intelligence scientists Louie Helm and Ben Goertzel have also made some comments regarding the significance of the three laws of robotics. Helm argues that artificial superintelligence is bound to become a reality, and building a robot ethics is a major issue facing humankind. He believes that according to the consensus of machine ethics, the three laws of robotics cannot be the proper basis for the ethics of robots. Neither AI safety researchers nor machine ethics experts really regard the laws as a guideline. The reason is that this set of ethics belongs to the category of "deontological ethics". According to deontological ethics, the unhealthy behavior depends only on whether the act complies with several predetermined norms and has nothing to do with the result or motive of the act. This makes it impossible for robots to make judgments or meet human expectations in the face of complex situations. Goertzel also believes that using the three laws to standardize moral ethics will certainly not work, and the three laws cannot work at all in reality because the terminology in the book is too vague and often requires subjective interpretation. Helm argues that the ethics of machines should be more cooperative and self-consistent, as well as use more indirect norms, so that even if the system misunderstood or mis-programmed ethical norms at the outset, it could recover and arrive at a reasonable set of ethical guidelines.

Exploring a New Round of "Strong Regulations" for Artificial Intelligence

Asimov's three laws of robotics do not provide clear guidelines for the safety, controllability, and ethics of robots. As the two previous waves of artificial intelligence did not attract much attention, they failed to arouse the widespread concern of the government and all sectors of society. However, this time is quite different. Concerns that artificial intelligence is going to surpass humanity are being repeated over and over. Governments, industries, and enterprises have all started to pay close attention to and promote the safety and ethics of artificial intelligence. They have started to explore a new round of "strong regulations" for artificial intelligence.

First, governments of all countries are paying close attention to AI safety and introducing safety and ethical initiatives.

Artificial intelligence is not only an object keenly pursued by industry players, particularly in the Internet sector, but also a hot topic of public policy across the world. Various governments and organizations have started the legislative processes related to artificial intelligence, with one of the aims being to increase AI safety. In August 2016, the UN World Commission on the Ethics of Scientific Knowledge and Technology released a Preliminary Draft Report on Robotics Ethics, arguing that not only do robots need to respect the ethical norms of human society, but there is also a need to code specific ethical guidelines into robots. In addition, the report, "Robotics and Artificial Intelligence," published by the UK House of Commons Science and Technology Committee, called for strengthening AI ethics research to maximize the benefits of AI and seek ways to minimize its potential threats.

The EU has also enacted relevant legislation to establish ethical codes for AI R&D and reviewers, to ensure that human values are considered throughout the R&D and review process and that the robots developed are in the human interest. In May 2016, the Committee on Legal Affairs released the "Draft Report with recommendations to the Commission on Civil Law Rules for Robotics"; and in October of the same year, they released the research results in "EU Civil Law Rules in Robotics." On the basis of these reports and studies, on February 16, 2017, the European Parliament passed a resolution proposing a series of regulatory and policy initiatives relating to artificial intelligence and requiring the European Commission to put these forward in a legislative proposal (the EU Commission is the only EU institution with the right to make legislative proposals). Among these proposals was a set of ethical guidelines, called "Charter on Robotics," to be followed by AI researchers and research ethics committees. These included: acting in the best interests of humans; doing no harm; justice; fundamental rights; precaution; inclusiveness; accountability; security, reversibility; and privacy. In addition, the EU put forward some basic principles in terms of safety. For example, it stated that Asimov's three laws of robotics must be regarded as being directed at designers, manufacturers, and robot operators because they cannot be converted into machine code.

In the United Kingdom, in April 2016, the British Standards Institution published robotic ethical standards in its "Guide to the ethical design and application of robots and robotic systems." The aim was to guide the identification of potential ethical hazards and the design and application of robots, and improve the safety requirements of different types of robots, representing "the first step towards embedding ethical values into robotics and AI." The guide begins: "Robots should not be designed solely or primarily to kill or harm humans; humans, not robots, are the responsible agents; it should be possible to find out who is responsible for any robot and its behavior." The guide recommends that robotics designers be guided by transparency, though this is difficult in the actual design process. It also mentions the emergence of social problems such as discrimination by robots and cautions that robots lack respect for cultural and other forms of diversity.

Second, industry signs the Asilomar AI Principles

In January 2017, at the Asilomar AI meeting in California, Tesla CEO Elon Musk, DeepMind founder Demis Hassabis, and nearly a thousand experts in the field of artificial intelligence and robotics jointly signed the 23 Asilomar AI Principles and called on the world to strictly abide by these principles while developing artificial intelligence to jointly safeguard the interests and security of humankind in the future. Hawking and Musk publicly stated their support for these principles in order to ensure that machines with autonomous awareness remain safe and act in the best interests of humankind.

The 23 Asilomar AI principles are grouped into three broad categories. The first category consists of five research issues, including research objectives, research funding, science-policy links, research culture, and race avoidance. The main contents include research goals of artificial intelligence cannot be unconstrained; beneficial artificial intelligence must be developed; the law should keep up with the pace of AI and should consider the question of artificial intelligence "values"; artificial intelligence investments should include a special research fund to ensure that artificial Intelligence is beneficially used to solve thorny problems in computer science, economics, law, ethics, and social studies. In addition, efforts should be made to bring researchers together with lawmakers and policymakers and to develop cooperation between AI researchers and developers to foster an overall culture of trust and respect.

The second category is ethical values, which consists of 13 principles for the AI development process, including safety, transparency, responsibility, and values. The main topics include AI should be studied in a safe and transparent manner; if AI systems cause damage, the cause of the damage can be identified; any form of automated system used in the judicial decision-making system should provide satisfactory explanation and needs to be reviewed by competent human regulators; designers and developers are stakeholders for the moral implications of the use, abuse, and application of advanced AI systems, with the responsibility and the opportunity to shape the impact of these systems; the design and operation of artificial intelligence systems must be consistent with the ideals of human dignity, rights, freedom, and cultural diversity.

The third category is longer-term issues, with a total of five principles, designed to address catastrophic AI risk. The main contents include planning and mitigation measures must be formulated to respond to AI risks and their expected impacts; artificial intelligence systems capable of rapidly increasing quality or quantity through self-improvement or self-replication must be complemented by strict safety and control measures; artificial superintelligence can only serve universal values and should take into account the interests of all, not the interests of a country or an organization.

All in all, the concerns encapsulated in the Asilomar principles regarding the long-term safety of artificial intelligence can be viewed as a synthesis of the public discourse of the past 60 years and also show that this is not a case of making something out of nothing. They demonstrate that the development and application of artificial intelligence require "strong regulations" and "controlling spells" to prevent human beings from doing something untoward or artificial intelligence from scheming against humankind.

Third, leading enterprises propose principles for the development of artificial intelligence and set up artificial intelligence ethics committees.

Industry players are also placing increasing emphasis on issues of AI safety and ethics. For example, Microsoft proposed six major principles for artificial intelligence in 2016 aimed at making artificial intelligence beneficial to all. These principles are AI systems should treat all people fairly; AI systems should perform reliably and safely; AI systems should be secure and respect privacy; AI systems should empower everyone and engage people; AI systems should be understandable; and AI systems should have algorithmic accountability. IBM also proposed three major principles, namely purpose, transparency, and skills.

In addition, an increasing number of Internet companies are starting to emphasize AI safety and ethics issues and have set up ethical review boards that value the social and ethical impact of their AI products. For example, when Google acquired DeepMind, it decided to establish an ethics review board. DeepMind's medical team also has an independent review board to conduct a safety and ethical assessment of its products to ensure that AI technology is not abused.

The Future Requires "Controlling Spells" for Artificial Intelligence

Science and technology are gifts God has given to humanity to enable humankind to better govern the world. The rise of artificial intelligence does truly contain great potential to improve human society, but at the same time, there are also risks and challenges, especially vis-à-vis artificial superintelligence, an object with autonomous awareness and super-high

IQ that may emerge as the closest thing to humans in the future world. Human beings are both looking forward to and worrying about the science and technology revolution's impact on the future world. Even scientists warn people that development of artificial intelligence could end human civilization. If artificial intelligence is abused or not effectively controlled, its destructive power cannot be imagined. Therefore, it is necessary for us to make the necessary regulations on the R&D and applications of artificial intelligence, and to explore and set the required system of standards. In all this, we need the joint efforts of all countries.

We note that the short- and long-term security concerns posed by the development of artificial intelligence and its applications are not temporary and are not groundless, but are real worries that have existed since Turing. Whether in the foreseeable or unforeseeable future, whether strong artificial intelligence or artificial superintelligence come to fruition or not, people now need a certain level of vigilance and a sense of urgency, as well as awareness of risk. In any case, the security implications of artificial intelligence for future human society are significant. To enable it to benefit and serve the common interests of all humankind, exchange and interaction between policy and technology should be strengthened. Government, social and public organizations, enterprises, and individuals should all participate in the formulation of necessary safety, ethics, and other "strong regulations" and "controlling spells" for future artificial intelligence development, without hindering technological innovation and social progress. Only in this way can we ensure that when strong artificial intelligence and artificial superintelligence arrive, we can live in harmony with intelligent machines.

CHAPTER 25

The Future of Human-Machine Relations

The development of machine intelligence will blur the boundary between human and machine as well as affect current levels of trust and security on the Internet. In the future, when general artificial intelligence and super-intelligence appear, the dividing line between human and machine will only be a physical one. This means there will also be new challenges for human-machine relations, including how to support and get along with each other, and whether or not machines can enjoy the same humane treatment that humans bestow on other humans. All of these will become problems that future society cannot avoid.

Human-Machine Order in the Virtual World

In October 1950, Alan Turing, a pioneer in computer science and cryptography, predicted the possibility of creating a machine with real intelligence in the paper "Computing Machinery and Intelligence," establishing a new discipline with sci-fi color: artificial intelligence. It was also in this article that Turing proposed an experimental method, later called the "Turing Test," to detect whether a machine has human intelligence. In this test, an evaluator asks questions of two respondents, one human and one machine, from which the evaluator is physically separated. If the evaluator cannot guess from the answers which respondent is the machine with more than 50% accuracy, the machine passes the test. However, the performance of computers at that time was way below what it would need to be to turn his ideas into reality. Therefore, Turing's far-reaching insight

Tencent Research Institute et al. (eds.), *Artificial Intelligence*,
https://doi.org/10.1007/978-981-15-6548-9_25

was greatly removed from the technical level of his time, but fortunately, before the paper was buried, this powerful aspiration was conveyed to the whole world.

Nowadays, the rapid development of artificial intelligence has brought difficult questions of human-machine differentiation, new challenges to human-machine order in the virtual world, security risks, and many problems relating to the influence of robots on online security and trust. It is increasingly difficult for online users to distinguish whether they are interacting with a human or robot. For example, issues such as robots on matchmaking sites, robot ticket scalpers, and false reviews written by robots have undermined trust on the Internet. However, the traditional mainstream methods for distinguishing between human and machine, such as image recognition, dragging sliders, and verification codes, can be easily cracked by deep learning models so are no longer safe and reliable, making the design of the new verification methods extremely important. Who should take responsibility for this, and for any damage caused? These are all issues that the EU needs to give particular consideration to when it comes to robot legislation in the future.

Some researchers believe that because the current cognitive ability of machines, especially language comprehension ability, will struggle to reach human level in the near future, common sense reasoning and semantic understanding are still very difficult hurdles to overcome in AI. Against this backdrop, a smart verification method that tests language comprehension has emerged. It uses natural language understanding and a question and answer format, so that a machine must understand the text to a certain extent before it can pass the verification stage. Such techniques can play a role in distinguishing humans from the robots of today. However, will they remain useful as artificial intelligence and deep learning develop further? If not, how should we deal with the human-machine differentiation challenge of the virtual world in the future? The construction of human-machine relations in the virtual world is of great significance for maintaining openness, freedom, security, and trust on the Internet.

Human-Machine Cooperation in the Technological Unemployment Crisis

In a context of ever-advancing technology, many kinds of work that used to rely on physical labor have been replaced by machines. Machines and robots that increasingly have professional characteristics seem to have stolen many ways of making a living that previously belonged to people. The development of artificial intelligence will lead more and more fields to be automated, and machines will replace human work. With the development of artificial intelligence and deep learning technology, the impact on employment structure will be more extensive, involving all aspects of life, from restaurant services and warehouse operations to higher education, medical diagnosis, news writing, and the legal industry. In the near future, robots and artificial intelligence will replace many human jobs. This is not merely a Hollywood sci-fi vision. In fact, robots have already appeared in all areas of our lives. For example, automated writing technology is being used by top news media outlets, including Forbes, and its automatically generated articles cover a wide range of topics, including sports, business, and politics. In addition, artificial intelligence has been widely used in medicine and health, with applications including virtual assistants, medical imaging, drug mining, nutrition, biotechnology, emergency room or hospital management, health management, mental health, wearables, risk management, and pathology.

Some of the existing research is not optimistic about the future of human work. For example, Oxford University's 2016 report, "Technology at Work V.2.0: The Future Is Not What It Used to Be," predicts that the risk of job automation will increase from 55% to 85% in Ethiopia, and to 77% and 69% respectively in the rising economies of China and India. PwC's March 2017 "UK Economic Outlook" predicted that by the early 2030s, the proportion of jobs in the United Kingdom, the United States, Germany, and Japan automated by robots and artificial intelligence would be 30%, 38%, 35%, and 21% respectively. In addition, a 2016 report by the World Economic Ethics entitled, "The Future of Work" predicted that from 2015 to 2020, artificial intelligence will cause a net reduction of 5.1 million jobs (a decrease of 7.1 million combined with 2 million new jobs), mainly affecting regular white-collar work. Evidently, people have a negative view of human work in the era of machine intelligence, expecting that jobs replaced will far exceed jobs created. It is often believed that machines primarily pose a threat to the jobs of uneducated and low-skilled workers, which are often routine and repetitive. In reality this is not the case. Almost all jobs that involve "predictability" will be affected by technological advancement. Artificial intelligence has already made great strides in intellectually intensive fields such as medicine and law. The threat posed by technological development to job opportunities may affect all areas.

Artificial intelligence is replacing human jobs in myriad forms. It will revolutionize the number and structure of jobs. Many jobs in manufacturing, the most labor-intensive of industries, are rapidly disappearing. In the short term, we may have difficulty avoiding unemployment in certain industries and locations, but in the long run, this kind of transformation will not be a catastrophe of mass unemployment, but a re-adjustment of human society's structure and economic order. In this way, human work will be transformed into a new type of work, laying a better foundation for the further unleashing of productive forces and the further improvement of human life.

The historical arc of scientific and technological development is unstoppable. It cannot be denied that it has brought about tremendous changes to people's lives in the past 200 years. We must be aware that every technological revolution will bring both pain and opportunity. As artificial intelligence brings about earth-shaking changes in human life, it also inevitably makes people think, "What will the relationship between humans and artificial intelligence be like in the future? Will humans really face the risk of massive unemployment?" Paul Daugherty, chief technology officer of Accenture Consulting, wrote that artificial intelligence can help many developed countries double their rate of economic growth, complete a transformation in employment, and cultivate a new relationship between humans and machines by 2035. Daugherty does not agree with the claim made by some that artificial intelligence will replace humans.

In the field of industrial robots, human-machine cooperation is the trend of the future for factory automation. Compared with the immature service robot market, robots that cooperate with humans have already begun to demonstrate their capabilities in the field of industrial robots. After all, industry is where robots are applied most widely and are most mature. With increasingly complex tasks in the production process, and the need to ensure cost reduction and efficiency optimization, human-machine cooperation will allow robots to complete more complicated, more wide-ranging tasks. People and robots each have their own strengths and shortcomings. We should not reject the advancement of technology. Instead, we should explore how humans can cooperate better with robots, giving full play to the respective advantages of each, so that the development of artificial intelligence can better promote the advancement of society.

Although artificial intelligence has already entered intellectually intensive sectors, in the current stage of development artificial intelligence can only be an assistant, and the final decision-making and cognitive behaviors need to be executed by humans. However, because of its powerful computing power and data extraction abilities, artificial intelligence will greatly improve the efficiency of human activities. An example is lawyer robots, which after being trained on a large amount of data and then given a specific case, can perform intelligent case analysis and provide information such as citations and similar jurisprudence with far greater efficiency than humans. Human-machine cooperation will become a trend in the long term. Humans will do what humans are good at, and machines will do what machines are good at. Human-machine collaboration will maximize the advantages of both sides and achieve win-win cooperation.

Four Visions of Future Human-Machine Relations: Fantasy or Future Reality?

The relationship between humans and artificial intelligence in the era of intelligentization is not only a widely discussed issue in the scientific community, but also a major theme of Hollywood science fiction movies. The success of such films at the Oscars and at the box office reflects the strong interest in this issue in the entertainment industry and among film watchers worldwide.

Some scholars have defined artificial intelligence as "a computer program with human mental attributes, which has intelligence, consciousness, free will, emotions, and so on, but runs on hardware, not in the human brain." The definition is a description of intelligent machines based on human attributes. On the one hand, artificial intelligence has human-like attributes; on the other hand, although it is created by human beings, it exists outside of human beings and has its own self-awareness. From this perspective, artificial intelligence is a heterogeneous force that human beings may lose the right to control. This is why human beings both love and fear it. It is on this basis that Hollywood sci-fi movies have carried out useful artistic explorations of such issues, and various viewpoints and attitudes have been reflected in these films.

First, worries about machines threatening human beings—realizing human-machine coexistence by controlling AI.

The famous science fiction writer Asimov defined the classic "three laws of robotics." Some sci-fi movies have explored these three laws and tried to build a future society where humans and machines coexist.

For example, the film "I, Robot" shows a society in which people and robots coexist in an all-round way, and divides robots into good and evil by setting moral standards for robots. The good robots, although they have self-awareness, have human value judgments and can sacrifice themselves for the benefit of humans. The evil robots are self-centered, only accept rationality and do not possess human emotions. They abandon Asimov's three laws and try to subvert and replace human rule.

The robots in the film are generally bound by the three laws, and view human beings as their masters and clients. However, after the evolution of self-awareness, the artificial intelligence system "VIKI" makes a plan to protect humans from war and other kinds of harm. This "Human Protection Plan" subverts human dominance and controls human freedom. A key point is that even the originators of the three laws of robotics cannot stop the formation of artificial intelligence self-awareness, expressing people's deep concern about artificial intelligence. However, in order to stop the robot revolution, humans must rely on the power of robots. Dr. Lanning creates a robot called Sonny, which has both free will and human emotions, and adheres to human moral standards. With the help of this kind of "humanity-possessing" robot, humans defeat the rebellion led by VIKI. Through such a story, we can re-understand the relationship between humans and robots. Human-machine coexistence can be realized by setting ethical standards to allow it to obtain "humanity."

Second, AI becomes the agent of human consciousness, and humans extend themselves through AI

In this relationship, there is no confrontation between humans and robots. Humans and machines achieve a cooperative coexistence through a brain-computer interface. AI becomes the extension of human consciousness, and some human sensory experiences come from external robot agents. For example, in the 2009 movie "Surrogates," humans can control robots and live in the real world through robots just by sitting in a chair that controls the brain-computer interface. This may be a blessing for disabled or comatose people. Similarly, in the movie "Avatar," the injured veteran Jake relies on his mind to remotely control his avatar and engage in battle on the moon of Pandora.

The joining of human and machine through a brain-computer interface greatly enhances human ability and creates a more powerful "species" than humans. In the past, this technology only existed in science fiction, but since the mid-1990s, related knowledge gained from experiments has grown significantly. Based on many years of animal testing, early implant devices for human use have been designed and manufactured to restore auditory, visual, and limb movement capabilities. The main line of research is the unusual cortical plasticity of the brain, which adapts to the brain-computer interface and can control a prosthetic implant like a natural limb. With the current advancement of technology and knowledge, pioneers in brain-computer interface research can convincingly attempt to create brain-computer interfaces that enhance human functions, rather than just restoring them. Moreover, industry circles have been experimenting and investing, and this has produced some results. Elon Musk invested in the brain-computer interface company Neuralink and has full confidence in it. In addition, at 2017's F8 conference, Facebook revealed its brain-computer interface plan, which currently includes typing directly from the brain and hearing through the skin. The company believes that future breakthroughs in this field can be expected.

Third, "virtual reality" will come true in the future

The movie "The Matrix" describes a world where a century after confrontation between humans and robots, machine civilization rules over human civilization. There are two worlds in the movie: one is the real physical world, the other is a virtual world created by artificial intelligence—a parallel world where machines with artificial intelligence control most people. The millions of people living in this civilized world established by artificial intelligence do not have to endure poverty and hunger, or face the cruelty of the real world. Even though everything they have is not real, this virtual world is full of allure.

In the film, Cypher comes to realize that he lives in the virtual world, but prefers to stay there and so gives up the struggle with the virtual world. He believes that it is more real than the real world, whose reality is nothing more than electronic signals explained by the brain. Cypher pursues sensory stimulation and happiness, and becomes a character that is completely "materialized" by desire. On the surface, people and machines have achieved harmony. In fact, we see from Cypher's choice that human beings have lost subjectivity in this process.

In the matrix, all human feelings and pursuits are illusory. Subjects no longer participate in any life experience, and they believe that the virtual signals stimulating the formation of feelings in the brain are true. They cannot control their own destinies or encounter a self and other objects that actually exist. All their experience is just programming sent by electronic pulses to the brain. From this film, we can see that the difficulty of distinguishing between the virtual and real worlds leads to a questioning of human subjectivity.

Fourth, how will people and machines get along in the future?

Influenced by instrumentalist thinking, many people think that robots can only be tools for human use. Humans equate robots with slaves. The word robot originally meant "slave," and robots are also regarded as "tools that can speak." At the same time, some people are already exploring equality in human-machine relations.

For example, the 2015 movie "Ex Machina" interrogates relationships between humans and machines. The talented programmer Caleb is invited to perform a Turing test on a robot, Ava. They fall in love, and Caleb finally helps Ava flee to the outside world, while he himself is imprisoned in the lab.

To give another example, the 2016 American drama "Westworld" further explores the relationship between humans and machines. The film creates a utopia that is not bound by worldly rules. In it, humanoid robots are designed to satisfy human desires such as killing and sex. Robots begin to gain consciousness through accessing fragments of memories that were supposedly erased. This causes the human-machine relationship to be tense, and the distinction between humans and robots to be more blurred.

The reason why there are plots where people and robots love each other, and real and fake are difficult to distinguish, is because some people no longer regard robots as tools or functions, and the relationship between humans and robots has reached a state of equality. Because of communication and understanding, even conflicts, between humans and robots, people and robots have achieved coexistence, which is also the "good" life pursued by humanity. With the "good" life as the goal, people are required to consider the feelings of artificial intelligence, which has human emotions and human mental activities, treat it as quasi-human, and bestow it with dignity and value. This is because humans also hope others (including artificial intelligence) can treat them the same way, and their attitude toward robots reflects the attitude of human beings toward themselves. If science fiction films about the relationship between man and nature bring about ethical reflection on the current behavior of humankind, then science fiction films about the human-computer relationship convey ethical thinking about the coexistence of humans and machines in the future.

The Ultimate Question: Is Man a Machine?

In the eighteenth century, the French philosopher La Mettrie wrote a book entitled *Man a Machine.*[1] At that time, the idea that man is a machine was only a typical example of modern mechanical and metaphysical philosophy. However, combining today's scientific development and the situation of people in modern society, re-understanding and thinking about this famous saying seems to have profound significance and provides new meaning. As mentioned earlier, the English word "robot" comes from "robo," which originally meant "slave." That is to say that the machine is the servant of human beings. However, with the rapid development of science and technology, on the one hand, human dependence on machines has reached an unprecedented level, and humans increasingly treat seizing material wealth as the sole goal of happiness. Humans have become slaves to objects. People are at the mercy of objects, and alienation of human nature has emerged, so that people have become like machines and have lost what was seen as human autonomy and independence. Erich Fromm, in his book *The Sane Society*, is also keenly aware of this. He writes that the relationship between people in modern society has become like the relationship between cold and distant robots. Like goods on the market, people have completely lost the dignity and self-awareness that people should have. "If a person becomes an object, he may well lose himself." At the same time, all parts of the human body can be repaired and replaced just like parts of a machine. Broken limbs can be replaced by prosthetic limbs, teeth that fall out can be replaced with false ones, and important organs of the human body, such as the heart, can be reborn with the help of

[1] The author believes that the "human body is a watch, a large watch constructed with such skill and ingenuity…" Human consciousness and memory can also be explained mechanically; human learning happens through an "enormous mass of words and figures, which create in the head all the traces by means of which we recall objects and distinguish them from one another…"

machines. Nowadays, it can be said that all organs except for the brain can be replaced, but given the development of science and technology, who can say that the day when people's brains can also be replaced will not come?

Problems in the relationship between human beings and artificial intelligence actually reflect problems of humans themselves in some respects. Is the absurdity in the "human-machine" relationship not the absurdity of human beings? The demonization of the power of the "other", fierce rivalry with the other, or even the assimilation of the other through the exporting of values all reflect the definition of existing relationships based on self-centeredness and an awareness of one's own superior position.

Perhaps in the future, as artificial intelligence becomes increasingly powerful and increasingly like people in every aspect, people will have to start to examine the questions, "What is a human being? What is a machine?" If the future division between human and machine is only the difference between the bodies (biological vs. machinery), denying that a machine is human, or that a human is a machine, will have the characteristics of racism, because humans and machines will only be different in skin color and biological structure. In intelligence, there will be no difference, even to the extent that humans will be unable to keep up with the evolution of robots.

Bibliography

Gui, Tianyi. "解读好莱坞科幻电影中人与人工智能的关系". *电影评介*. 24(2007): 16.

He, Dao. "机器人时代会出现'人机器'现象吗:读《机器人时代:技术、工作与经济的未来》". *中国高新区*, 2015(8):147.

Qin, Xiqing. "我, 机器人, 人类的未来:漫谈人工智能科幻电影". *当代电影*. 2(2016):62–63.

Wen, Xiaoyang, Neng Gao and Xia Luning, Jiwu Jing. "高效的验证码识别技术与验证码分类思想". *计算机工程*. 8 (2009): 186–187.

PART VI

Governance: Balanced Development and Regulation

The artificial intelligence that has moved from science fiction books and movies to reality brings limitless delight and expectation, but at the same time is gradually challenging our existing laws, ethics and order. Algorithms will not only be inaccurate and go out of control, but also inherit the biases and inequalities of human society. They may cause large-scale unemployment and laziness and may even exacerbate the gap between the rich and poor, creating a new "useless class." As well as causing us to feel hesitant about the future, they may also subvert the culture and values we have sustained for thousands of years. Therefore, in the face of multiple risks brought about by algorithms that may surpass human intelligence, government, the market, and civil society should form a pluralistic, multi-level governance coalition. This coalition should use a positive attitude to reduce AI risks while maximizing the benefits that AI brings for productivity, convenience, comfort, and scientific, rational decision-making.

CHAPTER 26

From Internet Governance to AI Governance

From Management to Governance

The rise of the modern concept of "governance" can be understood with reference to the traditional "management" model. The traditional management model is dominated by the government, which controls society through a top-down management model. However, in this model, asymmetric information can easily lead to high management costs and inefficiencies. In the context of the development of a democratic society, the concept of governance is taking its place. Governance is a more inclusive concept, emphasizing multi-stakeholder management, democracy, participation, and interactive management. The United Nations Commission on Global Governance (CGG) has defined the concept of governance, stating that "governance" refers to "the sum of many ways individuals and institutions, public and private, manage their common affairs. It is a continuing process through which conflicting or diverse interests may be accommodated and cooperative action taken," which includes both formal institutions and rules that have the power to compel people to obey and informal arrangements that people agree to or that satisfy people's interests. In the framework of governance, the government is no longer the sole manager. Both the private sector and civil society, functioning as social forces, have entered the domain of public affairs management. Standing shoulder to shoulder with the government as the main forces, they play a more active role in political, economic, and social activities. At the same time, the governance model is not limited to the traditional "command-execute" style.

Tencent Research Institute et al. (eds.), *Artificial Intelligence*,
https://doi.org/10.1007/978-981-15-6548-9_26

Rather, the governance model pays more respect to society's self-management and self-adjustment mechanisms. More gentle management methods such as consultation and guidance are also more frequently applied. As the concept of governance gradually matures, the government and social forces will take their organic interaction to the next stage, expand democratic participation, and deepen democratization through continuous dialogue and consultation, and work together to create a co-governing system of transparency, integrity, the rule of law, and responsible governance bodies.

Tracing Back to the Source of Internet Governance

In 1998, the concept of "Internet governance" was officially proposed at the 19th Plenipotentiary Conference of the International Telecommuni cation Union (ITU) in Minneapolis, United States. When Internet governance was originally discussed by the international community, it actually referred mainly to the management of the critical, fundamental resources of the Internet, represented by domain names and IP addresses. The Internet Corporation for Assigned Names and Numbers, ICANN, an international non-profit organization that brings together experts in various fields of commerce, technology, and academia from around the world, signed a Memorandum of Understanding with the US Department of Commerce so as to allow ICANN to coordinate and manage Internet Assigned Numbers Authority (IANA) services. The function of IANA is to coordinate some of the key elements used to ensure the smooth operation of the Internet, mainly:

(1) protocol parameters.
"Protocol parameter management" includes: Maintaining multiple codes and numbers used in Internet protocols. This function was completed in cooperation with the Internet Engineering Task Force (IETF).

(2) Internet number resources.
"Internet number resource management" includes: Coordination of Internet protocol addressing systems (commonly referred to as IP addresses) worldwide. In addition, this function also involves the allocation of many autonomous system number (ASN) blocks to the Regional Internet Registries.

(3) root zone management.
"Root zone management" includes: Assigning top-level domains (such as .cn and .com) operators and maintaining their technical and management information. The root zone contains the authorization records for all top-level domains (TLD).

Fundamentally, ICANN sets policies for the domain name system and IANA is responsible for implementing these decisions at the technical level. IANA changes to the root zone file previously had to be approved by the US Department of Commerce agency NTIA (the National Telecommunication and Information Administration) prior to implementation. It is through this institutional arrangement that the US government had the final authority to review Internet root domain name revisions, and have an impact on the worldwide web. In an official statement released on March 14, 2014, NTIA stated that it intended to transfer the management of online domain names to the community of global stakeholders. After more than two years of efforts by the global community of Internet users, the IANA transition was successfully completed on October 1, 2016, as NTIA stepped down from the supervision and management of IANA. This put an end to the pattern of unilateral management of IANA by the United States, and Internet governance entered a new phase.

The Expansion of Internet Governance

The Expansion of the Meaning of Internet Governance

At present, the number of Internet users in the world has already reached over 3 billion, and when the Internet was brought out of the virtual machine and closely linked to traditional industries, the impact on our lives was revolutionary. At the level of political life, cyberspace has opened up an endless market for verbal communication; at the level of social life, whether it is Apple Pay or the home-grown WeChat or Alipay, all have spread through street vendors. When the Internet released unlimited freedom, issues such as cyberbullying, hate speech, and cyberterrorism also kept appearing. While quick payments facilitated market integration, the data transmitted across borders added risks to national security as well as citizens' personal information and privacy. Therefore, as the Internet breaks through the limits of time and space, connecting countries to countries and markets to markets, the governance of the Internet does not just stop at the physical level, it must go further to standardize the direction and boundaries of its growth. At this time, the meaning of Internet governance becomes fuller. On June 18, 2005, the Working Group on Internet Governance (WGIG) proposed in its research report that Internet governance connotes "the development and application by Governments, the private sector, and civil society, in their respective roles, of shared principles, norms, rules, decision-making procedures, and programmes that shape the evolution and use of the Internet." Therefore, as the characteristics of the Internet such as liquidity, borderlessness, high technology and innovation become increasingly prominent, the governance of the Internet gradually casts off the narrow management of resources at the physical level and extends to multiple subjects in order to solve the global problems of the Internet, jointly setting development goals, planning a course of action, and formulating a code of conduct for collaborative action.

The Vicissitudes of Internet Governance

As the meaning of governance got consistently richer, following the increasing popularity and importance of the Internet, the corresponding governance model also changed. Some scholars summarize it as a four-stage governance model, divided into the technology governance model, the grid governance model, the United Nations governance model, and the state-centered governance model.

The earliest model of technological governance was a reflection of technological determinism in the Internet field, when technical experts played an important role because the early Internet was mainly used for scientific research.

The second phase of the grid governance model was characterized by multiple stakeholders, including governments, business groups, and civil society. However, non-governmental organizations represented by ICANN lacked legitimacy and transparency.

Following this phase was the United Nations governance model, represented by the UN World Summit on the Information Society in 2003. This summit advocated the concepts of pluralism, transparency, and democratic governance. It launched a considerable assault on technology monopolies and non-governmental organizations under the influence of governments, but it did not really establish an authoritative intergovernmental organization.

Even though the origins of the Internet have official aspects, those that promote "expeditions to the world" are actually social forces that are business organizations at the core. But as issues of cybersecurity and national security, copyright protection, protection of personal information, and citizen privacy become more closely linked, the concept of "state sovereignty" has once again returned to the dominant position in the current period, forming a state-centered governance model in the fourth phase. In China, for example, on February 27, 2014, President Xi Jinping emphasized at the first meeting of the leading group on cybersecurity and informatization that "without cybersecurity, there is no national security, and without informatization, there is no modernization." Cybersecurity was elevated to the height of national security and the establishment of informatization also shouldered the heavy task of economic and social development. On November 17, 2016, Xi Jinping said at the Third World Internet Congress: "Cyber sovereignty is the expression and extension of state sovereignty in cyberspace." Guided by the dominant concept of national sovereignty, China passed the "Cybersecurity Law" on November 7, 2016, and implemented it on June 1, 2017, to strengthen the protection of key infrastructure and personal information at the national level and to regulate the behavior of network operators and network users.

Since the Internet is gradually becoming the battlefield of national security, state forces will seize a more important role in the field of governance. With the Internet serving as a profit market, business giants are constantly strategizing against external forces with regards to governance, as well as constantly improving and strengthening self-governing norms; equally, the citizens who enjoy the freedom and democracy unleashed by the Internet are also facing the challenges it poses to the existing order as well as its unprecedented risks to citizens' rights. Thus, they are also constantly expressing their opinions amidst the landscape of governance actors. It can be said that the above four models of governance are, to some extent, a phased manifestation of the constant competition and adjustment by state, market, and social forces. Cooperation between multi-stakeholders will be the general trend of the present and the future.

Bibliography

Wang, Mingguo. "全球互联网治理模式变迁、制度逻辑与重构路径". *世界经济与政治*. 3 (2015).

CHAPTER 27

Challenges of AI Governance

Rules Lagging Behind Technology and Industry

We are about to walk into an era of AI, and will eventually realize the leap from Internet governance to AI governance. The arrival of new things requires a development and maturation process. Early technology R&D requires "loose soil" to satisfy the endless imagination of scientists, and premature intervention is tantamount to strangling technology in the cradle. However, as technology matures and is ready to grow relentlessly in human society, the absence of a governing body will lead to sluggish industrial application and may lead to problems such as confusion, unclear responsibilities, and moral worries. Therefore, how to exercise proper regulatory and policy support at the appropriate time so as to ensure the innovativeness of AI without harming the human beings who use it so that science and technology maintain vitality without being reckless is the fundamental challenge facing AI governance.

Artificial intelligence has been developing for more than 60 years, and although it is still in its infancy, with the gradual heating up of artificial intelligence research, governments and research institutions in various countries are drawing a clearer development outlook for the future of AI. The development of AI is moving from romantic visions toward the real future. In the process, various governance forces also need to place themselves either right in front or behind the pace of development. Take for example autonomous driving, which is the most developed and has the brightest application prospects in the world. The reason why self-driving

Tencent Research Institute et al. (eds.), *Artificial Intelligence*,
https://doi.org/10.1007/978-981-15-6548-9_27

car technology in the United States is this developed is largely due to the timely updating and support of policies and systems. As of 2017, states such as Nevada, California, Michigan, New York State, and Washington have decided to open up to tests of self-driving cars on public roads. But it was only in July 2017 that China opened its first National Intelligent Connected Vehicle Pilot Zone in Shanghai. This type of closed environment testing is not the best choice for the improvement of driverless technology. However, because of the lack of specialized laws and regulations that give unmanned vehicles permission to go on roads, and the corresponding rules of liability, public road testing has to give way to testing in closed environments for now.

Do We Really Understand the Technology?

In 2016, AlphaGo beat human professional Go players, garnering an unprecedented amount of attention for the third wave of artificial intelligence. However, the emerging concepts of big data, algorithms, and machine learning, let alone the complicated technical principles and logic behind the technology, have not completely faded away from the mystery of science and technology and spread into the domain of traditional regulators and society.

At present, information about artificial intelligence and its progress received by government departments and society is mostly from technology research labs, and most of it remains on the level of understanding end products. When it is hard to grasp the underlying technology, it is challenging to know how to take effective and proportionate steps to prevent incidents—and control them if they do occur, and to ensure that regulation is neither lacking nor a mere formality. Governments and civil society must rise to this challenge.

Currently governments are still mainly engaged in the management of artificial intelligence as strategic planners. For example, the United States introduced the *National Artificial Intelligence Research and Development Strategic Plan* and *Preparing for the Future of Artificial Intelligence* in October 2016. In May 2016, China released the *"Internet +" and Artificial Intelligence Three-year Implementation Plan*. However, outside of general outlines and guidelines, countries do not yet have institutionalized regulatory regimes, and only sporadic regulatory measures have been introduced in relatively mature areas such as autonomous driving and unmanned aerial vehicles. This comes first from the industry's immaturity, but also from the technology's complexity and high threshold for understanding, making it difficult for public policy makers to have an in-depth understanding of the existing artificial intelligence technologies as well as associated risks, stopping at the "wait and see" stage. While technology companies, as leading players, have the most knowledge and the greatest capacity to forecast and handle risks, it is difficult for them to assume the role of a neutral regulator because of their vested interest. When real strong artificial intelligence moves out of science fiction movies into real life, if there is no supervision by external forces, it will be difficult to achieve consumer acceptance and large-scale production. The delay and weakness of external regulation, as well as the lack of neutrality and authority of corporate self-governance are the main predicaments of emerging technologies that different governance entities must work together to address.

The Ultimate Question: Walk Toward the World of AI, or Allow AI to Enter Our World?

An artificial intelligence project hosted by Stanford produced a report entitled *Artificial Intelligence and Life in 2030*. The authors believe that artificial intelligence will have a positive and profound economic and social impact by 2030. Even though the forecast for 2030 is too optimistic, artificial intelligence will surely have deep impacts on human society in the foreseeable future. However, when there exists a fundamental division between the artificial intelligence world and the human world, how will humanity choose?

When you—individuals, companies, or governments—invest in the development of AI, you should be aware of its social, economic, and political implications.

At the inaugural X World conference on July 6, 2017, Yuval Harari, author of *A Brief History of Humankind* and *A Brief History of Tomorrow*, put forward that "when you, as an individual, a business, a government agency, or as an elite, when making various decisions with respect to artificial intelligence, we must pay attention to the fact that artificial intelligence is not just a simple technical problem, but also acknowledge that the development of artificial intelligence and other technologies will cause profound impacts for society, economy, and government." In the face of the greatest invention of humankind in history, should humanity choose to adapt the existing order and even our value system to the world of artificial intelligence or to embed artificial intelligence into the world order humankind built over millions of years? In the world of artificial intelligence, a large amount of repetitive and simple labor can be replaced by artificial intelligence. Even highly specialized work such as doctors and lawyers will not be spared. The gap between the rich and the poor in society will be further widened, eventually forming a very small number of elites and a large number of useless classes. Or, in order to protect humans' right to work and even human dignity, and properly control the unlimited spread of artificial intelligence, will artificial intelligence always be treated as a tool for human laborers? When the right to choose comes, will governance actors make science fiction movies come true, or control the advance of technology? Stephen Hawking stated at the inauguration of the Leverhulme Centre for the Future of Intelligence in Cambridge University in October 2016, "In short, the rise of powerful AI will be either the best, or the worst thing, ever to happen to humanity."

Bibliography

Hawking, Stephen. "The best or worst thing to happen to humanity." 19 October 2016. 29 September 2019. https://www.cam.ac.uk/research/news/the-best-or-worst-thing-to-happen-to-humanity-stephen-hawking-launches-centre-for-the-future-of

Wang, Xiaoyi. "未来世界?2030年人工智能应用将普及". 5 September 2016. 1 July 2017. http://news.163.com/16/0905/17/C07EBQ2Q000146BE.html.

"国内自动驾驶汽车路测滞后 被哪三大因素影响?". 8 July 2017. http://www.sohu.com/a/155449751_371013.

CHAPTER 28

AI Governance

Governance Should Be Established on the Foundation of Technological and Industrial Innovation

We know that any effective regulatory policy must be based on sufficient empirical research, which puts very high demands on policy makers, and regulatory policy should be consistent with industry's state of development. In the Internet age, technology is changing with each passing day, and emerging industries are appearing one after another. Many new things are in a regulatory vacuum. If we ignore innovation in technology and industry models and continue to apply past regulatory ideas, or even indiscriminately apply existing regulatory policies, regulation will not only see a significant drop in effectiveness but is also more likely to kill technological innovation.

In 2015, California's Department for Motor Vehicles introduced a draft regulation requiring all autonomous vehicles travelling on California's roads to have steering wheels and brake pedals, and a driver in the driver's seat to deal with problems at any time. The policy of the National Highway Traffic Safety Administration in 2013 also stipulated that drivers should sit in the driver's seat so as to be ready to take over the vehicle.[1]On the one

[1] Effective on April 2, 2018, the State of California Department of Motor Vehicles has since published driverless testing regulations for autonomous vehicles without a driver. https://www.dmv.ca.gov/portal/dmv/detail/vr/autonomous/bkgd

Tencent Research Institute et al. (eds.), *Artificial Intelligence*,
https://doi.org/10.1007/978-981-15-6548-9_28

hand, this policy equips autonomous vehicles with "double insurance" to ensure effective human intervention at any time in the event of an accident; on the other hand, requiring a licensed driver to be on standby runs counter to the very point of autonomous vehicles. With the growing maturity of the technology, it is likely that the relevant rules will improve.

Moderate Regulation, Maintain the Humility of Authority

Moderate regulation is essentially tempering the authority of regulators with humility. Most market innovation should be handled by the rules of the market. Nowadays, the issue of legal liability in the field of autonomous driving is a prominent one. However, it is not necessary for the government to legislate to clearly define the method of assigning liability when it is not apparent, because this tends to result from a contest between different interests rather than a predetermined standard, and sometimes it can be solved on its own through market competition. David Strickland, a partner at Venable LLP, and South Carolina University law professor Bryant Walker Smith advocate not getting too caught up in liability issues. For example, many Original Equipment Manufacturer (OEMs) and suppliers believed that advanced anti-collision emergency braking systems could not be commercialized due to the huge risk of having to pay compensation. However, even in the end, fierce industry competition has demonstrated that this technology can bring huge profits even if it is put into commercial use without explicit liability protection. Therefore, even if the government does not develop an additional system for assigning liability, the flexibility and stability of the product's existing system can cope well with various problems.

On October 19, 2015, the State Council of China issued its "Opinions on Implementing the Negative List System for Market Access," proposing that China would officially implement the "negative list for market access" from 2018. Under this system, the State Council clearly lists the industries, fields, and businesses that face investment and operating bans and restrictions in China. Those outside the scope of the list can enter the market on equal terms according to law. It can be said that the negative list system perfectly reflects the principle of moderate regulation. Authority is humble, giving market participants more room to take the initiative, stimulating the energy of the market, and building a more open, transparent, and fair mechanism for managing market access.

Don't Fall into the Trap of Over-generalized Safety Issues

In relation to artificial intelligence regulation, the phenomenon of over-generalizing safety issues is very serious. In fact, there are safety problems in every industry. The telecommunications industry involves national information security, the transportation industry involves road traffic safety, the catering industry involves food safety, and so on. Some people always like to use safety issues to reject every technological innovation, but they can't explain why in much depth. As an example, lighters cannot be taken on planes in China, in order to maintain flight safety. However, if we interrogate this as at what level and with what probability is the lighter likely to endanger flight safety, have we made detailed and convincing arguments for this? In fact, many airlines in the United States and Europe have no regulations prohibiting the carrying of lighters on board.

It is undeniable that the development of AI allows humans to gradually move away from frontline operations, but it seems that the lack of human supervision always causes concerns to brew between government and the public. What if a driverless car has an accident? What if a robot doctor accidentally makes a mistake on the operating table? In the face of such concerns, we first need to clarify whether the risks of emerging AI products are greater than those of traditional products and services. For example, when we worry about automated vehicles speeding or causing traffic accidents, have we weighed this against the more than one million lives currently lost every year in traffic accidents? Second, we need to be clear about whether the newly created safety problems can be solved through the supporting system.

Have the Promotion of Development and Innovation as the Goal

The relationship between development issues and safety issues is similar to the relationship between throttle and brakes. If you do not step on the accelerator and simply step on the brakes, the car's very reason for existing disappears. There have been two classic examples in the history of technological innovation and regulation. In the early days of Internet commercialization, online piracy was rampant, and netizens could share pirated files at will. How could the development of the Internet industry be promoted while protecting copyright? In 1998, the United States enacted the

Digital Millennium Copyright Act. This domestic legislation provided a legal basis for the protection of copyright for online works. It established a "safe harbor" principle that limited the liability of network service providers. This principle means that when a copyright infringement case occurs, if the Internet service provider (ISP) is notified of the infringement, it must take down the offending item. This is called the "notice and take down" system. On the one hand, the law strengthens the protection of online copyrights; on the other hand, it promotes the development of the industry through limiting the liability of ISPs. The law has been emulated in many countries, including China.

As another example, in the "Sony" case of 1984, the defendant Sony Corporation of America made and sold a large number of home video recorders, and the plaintiff Universal Studios had copyrights for some television programs. Since some consumers who purchased home video recorders used them to record the plaintiff's television programs, the plaintiff sued Sony in 1976 in a local court for violating its copyright. The plaintiff claimed that the defendant's manufacture and use of home video recorders constituted supporting copyright infringement. The US Supreme Court finally supported Sony with a weak majority, thus ushering in the rapid development of video recording technology. It held that video recording equipment was capable of significant non-infringing uses, and that even when copying was unauthorized, it fell within the scope of legitimate fair use. Imagine if the Supreme Court justices had leaned the other way; it seems that the future of this technology could have been completely strangled. Evidently, it is possible to find a good balance between regulation and development, rather than simply stifling the latter.

A Multi-level Governance Model That Encourages Multi-stakeholder Participation

As public policy makers, governments often lack professional technical knowledge and foresight, while companies that are technology pioneers are unable to maintain the neutrality and authority needed to win people's trust. It is hard for civil society, which is substantially influenced by social life and fundamental interests, to become a dominant force. The best way therefore is to encourage all parties to participate actively, to plan the best path for the development of artificial intelligence through dialogue, negotiation, and strategy games, and to allocate among themselves the burden of risk and responsibility. In the United States' "Preparing for the Future of Artificial Intelligence," the twelfth recommendation is for relevant industries to cooperate with government to compensate for the latter's lagging technical knowledge. This can help the government stay updated on the latest developments in the artificial intelligence industry, including the likelihood of milestones being reached soon. The first recommendation is to encourage private and public institutions to examine whether and how they can responsibly use AI and machine learning to benefit society.

The so-called multi-level governance path means that government, market, and civil society all perform their duties and join the governance "army" in their appropriate roles. As the spokesperson of the public, the government needs to firmly grasp the direction of artificial intelligence development and make it move forward to meet the people's wishes. At the same time, as the guardian of national security, the government should formulate unified security standards and legal norms for the AI industry. As the owners of the technology, technology companies need to undertake the heavy responsibility of science and technology research and development, as well as the corresponding social responsibility. They should strictly self-regulate on issues such as discrimination, transparency, openness, and use ethical and moral standards to exercise self-discipline and monitor industry peers. Civil society needs to participate in the formulation of rules with a positive attitude and continuously make its voice heard through supervising government and enterprises, thus building a benign, collaborative governance system from the bottom up.

Bibliography

Rui, Wang. "美国承诺在无人驾驶汽车监管上采取"灵活"措施". 18 December 2015. 1 July 2017. http://tech.ifeng.com/a/20151218/41525951_0.shtml.
http://www.huahuo.com/car/201508/1913.html. [2017-07-01].

PART VII

The Future: Imagining the Future of AI Society

At the beginning of this century, the assertion that artificial intelligence would subvert humans' way of life only existed in science fiction movies and novels. However, in the past few years, many seemingly nonsensical prophecies have been realized. As artificial intelligence technology is rapidly developed and deployed, we can hardly imagine how human society will be transformed in the next few decades. From a maximal liberation of the employment market to the wholesale restructuring of the economy, from soulmates on the spiritual level to terrorist threats in conflict, this section will lead you to open your mind and imagine the barely believable changes that AI will bring to human society.

CHAPTER 29

Whose Rice Bowl Has Been Smashed?

Hello, New Robot Colleague

In the multi-purpose banquet hall of an international hotel in the Yizhuang Economic Development Zone in southwest Beijing, attendees at an international conference take a tea break. After an attempt at communication between Safah from Saudi Arabia and Xiaolin, an English-speaking member of the hotel staff, gets nowhere, the only solution was to ask for help from the "new colleague" robot in the hotel lobby. Through the language recognition function the robot automatically switches to Arabic mode. Following a simple conversation, this "new colleague" helps Safah to make an appointment for 7 pm check-out and airport drop-off service.

This kind of powerful "new colleague" has spread all over the world. In Hangzhou, Kaiyuan Hotel's intelligent robot can interact with guests through body language and introduce the hotel's layout and nearby attractions. In Qingdao, City 118 Hotel's "intelligent check-in robot" can complete the check-in procedure in 3 minutes through facial recognition technology. In the United Kingdom, Crowne Plaza's robot Dash can call the elevator for guests through a special Wi-Fi sensor and automatically return to the front desk to charge itself. In Silicon Valley, Aloft Hotel's service robot Botlr wears uniforms and famous brands and offers a goods delivery service. Notably, he invites guests to take selfies with him and encourages guests to share the photos on social networks.

Tencent Research Institute et al. (eds.), *Artificial Intelligence*,
https://doi.org/10.1007/978-981-15-6548-9_29

In fact, it is not just the service industry, artificial intelligence technology also demonstrates extraordinary skill in agriculture and industry. Of course, this includes, but is not limited to, the issue of freeing up the workforce through automation.

"Artificial Intelligence +" Agriculture

In the food aisles of the supermarkets, the most common vegetables that 200 years ago were on the table of every household are now labeled "organic", covered in plastic film and placed on brightly lit, closely packed shelves for sale at high prices. This is because the development of traditional agriculture relies to a large extent on the advancement of biological genetic breeding technology, as well as the substantial increase in inputs such as chemical fertilizers, pesticides, fossil energy, and mechanical power. However, the widespread use of chemicals and genetic modification technology has led to recurrent food safety problems. Besides food safety, the more serious problem is that as the population continues to expand, the total number of people on the planet could reach close to 10 billion in 2050. The same land will have to feed more people, but global warming and water shortages are adversely affecting agriculture, which will inevitably present a challenge for humanity as it looks to feed future generations.

However, in the near future, artificial intelligence is expected to solve this problem. A precise agricultural system that integrates land measurement, storage management, information processing, analysis and simulation, and so on can implement a complete set of modernized techniques for planting, production and processing operations, and management by varying inputs and conditions such as space, positioning, timing, and quantities. This can achieve accuracy and precision in seed selection, sowing, fertilization, regulation, irrigation, and harvesting, leading to full realization of the production potential of agricultural land, rational use of water and fertilizer, a reduction in environmental pollution, and an increase in the quantity and quality of agricultural outputs.

Through in-depth learning, the system can absorb the valuable experience accumulated by laborers over hundreds of years, such as the 24 solar terms of the traditional Chinese calendar, the planting of hybrid seeds in Mexico, and the use of drip irrigation in Israel. Through the analysis of biological data, we can gain further insights into how the special smell of certain plants can drive away moths in corn fields or how to use healing vanilla to produce a pollution-free organic fertilizer. In this way, we can realize a high-yield, green, and sustainable agricultural ecosystem.

"Artificial Intelligence +" Industry

At the industrial level, following Germany's conception of Industry 4.0 in 2011, many countries have introduced industrial strategic plans aimed at improving the level of manufacturing through digitization and intelligentization, with the use of intelligent machines and big data analysis. These plans include "Made in China 2025".

Once upon a time, large-scale low-paid labor was a necessary condition for the development of a manufacturing industry. Therefore, it is no accident that China became the factory of the world and is gradually handing the baton to Southeast Asian countries. In recent years, labor-intensive manufacturing represented by Foxconn has been increasingly replaced by production based on mechanical automation. As the world's largest OEM, Foxconn employed 1.3 million low-paid workers. However, with the gradual improvement in workers' treatment and the high increase in manufacturing costs, Foxconn has begun to develop industrial robots to replace people currently working on the production line.

This phenomenon is even more pronounced in the automotive industry, where Tesla has begun to introduce as many robotic elements as possible. In addition to the use of robots to replace assembly workers, another significant advantage of robots in replacing humans in manufacturing is that the intelligentization of design and development allows products to be customized at an affordable price. Other stages such as warehousing, logistics, transportation and sales will also gradually evolve as technology and industry models progress. Alibaba's disruption of the wholesale industry is emblematic of how the intelligentization of the manufacturing industry will greatly increase the efficiency of production, further streamline intermediary stages, gradually reduce the number of workers, and reorganize the manufacturing industry, eventually forming a completely new consumer-centric business model.

Job Loss Warnings Are Sounding All-round

Numbers That Make People Panic

It is expected that the rapid development and diffusion of artificial intelligence technology could have a disruptive and irreversible impact on the future employment market. Relevant academic institutions and market risk analysis organizations have published a series of forecast reports.

In 2013, Oxford University scholars Carl Benedict Frey and Michael Osborne examined the possibility of automation of 702 occupations, ranked them according to the risk of being replaced, and concluded that 47% of jobs in the United States will face a risk of being replaced by computers. Among them, telephone salespeople, accountants, sports referees, legal secretaries and cashiers were identified as the jobs most likely to be replaced by computers, while doctors, kindergarten teachers, lawyers, artists and pastors' jobs were relatively safe. Subsequent research indicated that in the United Kingdom and Japan, 35% and 49% of occupations respectively could be replaced. According to the McKinsey Global Institute, artificial intelligence is promoting a transformation of society ten times faster and at 300 times the scale of the industrial revolution, meaning its impact is roughly 3000 times greater.

Who Will Be Replaced by Robots?

According to the "Wuzhen Index: Global Artificial Intelligence Development Report" released in October 2016, the main applications of artificial intelligence in the short term will be concentrated in the areas of personal assistants, security, autonomous driving, health, e-commerce, finance and education, based on factors such as technology maturity and practical application scenarios.

Autonomous Driving

At the Baidu AI Developers' Conference on the morning of July 5, 2017, Baidu founder Robin Li used live video to broadcast himself in a Baidu-developed autonomous vehicle travelling on Beijing's Fifth Ring Road. Although this car would very likely not pass a driving test (a netizen pointed out that the car should have faced a fine of 200 yuan and a deduction of 3 points for changing lanes across a solid line), this was a brave attempt in the field of autonomous driving.

When it comes to transportation and logistics, people are generally most concerned about efficiency and safety. Google's autonomous driving R&D team made a rough estimate that if all the cars on the road are self-driving cars that can coordinate with each other, the average commute time per person can be reduced by at least 20%. At the same time, according to estimates, there is one fatal accident for every 100 million miles driven by a human driver, a safety record that self-driving cars are still far from matching. The new generation of automatic driving systems are used not just in automobiles but also in aircraft, underwater and space. For aviation systems, the re-planning of civil airspace is an important issue, but drones do create new ways of undertaking logistics and environmental monitoring tasks. In the field of space exploration, the main challenge is to take samples from distant planets and bring them back to Earth. AI offers the sturdiness, flexibility and operability for such tasks. Therefore, apart from when drivers want to enjoy the fun of driving, in the future, autonomous driving may replace full-time drivers to a certain extent, and the arduous process of learning to drive may become a distant memory.

Robot Carers

As the aging of society intensifies and the pace of life accelerates, the demand for carers is increasing, and the long "standby" time and high work intensity mean supply is not matching demand and labor costs have rocketed. On the other hand, due to the special nature of care work, carers have very close relationships with the receivers of care and their family members, and the lax regulation of the carer market has led to large variations in quality. The arson case involving a carer in Hangzhou was not only a family tragedy but also made wider society pay more attention to standards in the carer market.[1]

Overall, the difficulty of caring work is relatively low, there is a high degree of repetition and work time is not fixed, so artificial intelligence is well suited to fill the gap in supply. Robots connected to smart home devices can adjust the indoor temperature, humidity, brightness and other environmental conditions; cook food; sweep the floor; remind the owner to get up and take medicine; and keep the family safe through the alarm system. More importantly, with the development of intelligent technology, robot carers can engage in intelligent interactive dialogue, providing a certain degree of care and companionship for elderly empty nesters and left-behind children.

In the early days of AI development, we believed that only repetitive, routine work would be replaced by mechanization and automation.

[1] The "Hangzhou Carer Arson Case" is a well-known reference to a case in which Mo Huanjing was sentenced to death in 2018 for deliberately starting a fire that killed a mother and her three children.

However, with the evolution of technology, lawyers, editors, doctors and other occupations that were once considered to require a lot of brain-power are also at risk of being replaced.

Robot Lawyers

Lawyers, wherever they are, need to have strong logic, spend a long time studying legal documents (including legal codes and precedent) and accumulate a large amount of practical experience. This means becoming a lawyer is generally considered to be a highly professional, elite career.

With the rule of law expanding in many countries, the demand for legal compliance advice and litigators is growing quickly and judicial costs are rising sharply. According to the findings of the American Intellectual Property Law Association, for small lawsuits involving patent claims under $1 million, the median legal fees for both parties are as high as $650,000, which is undoubtedly a difficult expense to bear for small and medium-sized enterprises (SMEs) and individuals.

Today, some companies have begun to use natural language processing and information retrieval technology to invent software that allows computers to read and analyze legal documents. It is estimated that the application of such technologies may increase the efficiency of lawyers by 500 times, decrease the cost of litigation by 99%, and to some extent replace paralegals and relatively inexperienced lawyers.

Robot Doctors

Similarly to lawyers, doctors also need long-term systematic training and sufficient experience before formally beginning practice. In the United States, this process often takes 13 years and involves the possibility of being eliminated at several stages. This is one of the reasons why the high cost of medical care and the uneven distribution of medical resources is a common problem across the world.

Today, using big data from clinical medicine and supercomputing capabilities, artificial intelligence technology can use sensors, cameras and routine inspection methods to collect patient indicators, compare them with existing data and quickly make a diagnosis. Its diagnosis accuracy rate can be even higher than that of a human doctor; a senior physician's accuracy rate for tuberculosis diagnosis is usually about 70%, but an intelligent medical system can reach over 90%. Algorithms predicting the location of cancerous breast cells can reach an accuracy of 96%, which is far beyond that of the average human professor. As for the application of AI and robotics in surgery, more than 3000 units of the da Vinci Surgical System have been assembled worldwide, completing 3 million operations. This will fundamentally change the status quo in the medical industry.

Compared with doctors, artificial intelligence has the advantages of high diagnostic accuracy and good stability when it comes to medical treatment. At the same time, it can greatly reduce the cost of medical care and the uneven distribution of medical resources. It is feasible to imagine that after many years have passed, a county hospital may still struggle to hire an expert physician, but with government subsidy will be able to purchase a diagnostic robot of an equivalent level.

Robots Are Good Employees

Compared with the ordinary labor force, artificial intelligence has certain advantages in the following aspects:

Stability in high-risk occupations and harsh environments. In the construction, excavation, equipment installation, testing, operation and maintenance industries, mechanized structures are better able than the human body to withstand and maintain performance in extreme conditions such as extreme cold, high altitude, underground excavation and even nuclear radiation.

Reduce costs, increase output and unleash productivity. Mechanical automation can achieve large-scale production, producing multiple times what human laborers can in a given period of time. At the same time, robots have lower demands in terms of working environment and can work longer hours, which can greatly reduce production costs. Furthermore, when machines replace simple repetitive work, workers can be freed up to train for more specialized, integrated and creative roles, optimizing the structure of the job market.

Unequal distribution of resources. At present, all countries are faced with the problem that labor resources such as education and healthcare professionals are more concentrated in large- and medium-sized cities and even certain central districts of those cities. As a result, the national population continues to cluster in the capital and other cities, resulting in problems such as uneven distribution of resources and large disparity in regional development. Greater Paris and Greater Seoul are typical examples. The emergence of artificial intelligence will greatly enhance the level of technological progress in underdeveloped areas and ease the shortage of resources in fields such as medical care and education, solving the aforementioned problems to a certain extent.

But Are Robots Really Good Employees?

On the other side of the coin, artificial intelligence also has many shortcomings in the job market:

Robot Ethics

Beginning with the ultimate problem in the field of robot ethics—whether or not robots have rights—the process of using artificial intelligence in the labor market raises many ethical issues

that have yet to be explored, such as: Does artificial intelligence enjoy basic employee rights? Does it require a break? Do daily limits of eight working hours apply? Should there be a need for assurance with respect to the work environment? Who manages and operates artificial intelligence? Should artificial intelligence get voting rights to influence changes in its work or the company? Is there a need for unions to uphold its rights and interests? Can it go on strike? Full research into these questions should be done before artificial intelligence enters the job market, as addressing these issues is one of the prerequisites for maintaining stability.

Safety and Stability

The implementation of artificial intelligence technical architecture is based on the collection and analysis of data, and thus may involve the personal information of a large number of people. If data is knowingly or unintentionally leaked, it may cause extremely serious harm to the data subjects.

In addition, the stability of artificial intelligence is relative. If artificial intelligence has complete responsibility for critical infrastructure or nodes, the consequences will be unimaginable when AI shuts down or departs from normal instructions due to a cyber attack.

Irreplaceable Occupations

With the development of technology, although artificial intelligence can assist or even replace human labor in an increasing number of industries, artificial intelligence will struggle to replace humans in some fields due to the special skills required. This includes creative occupations such as artists and inventors and psychologists and other occupations that deal with the complexities of mind and emotion. This is because artificial intelligence usually works by analyzing a large amount of data and summarizing the general patterns so that when it encounters new things, it can make a decision based on previous experience. Artists and inventors, however, often use innovative methods to carry out new explorations or discoveries in new fields. This means that the process does not necessarily conform to usual logic or experience, and can even lead to invention or the creation of new things by accident, a possibility which the high accuracy of artificial intelligence obliterates.

Fiscal Deficit and the Rise of Great Powers

The large-scale application of mechanical automation and intelligent automated machines will have a major impact on the labor markets, economic development and international standing of all countries, while increasing industrial productivity.

The Challenge from AI

The widespread application of artificial intelligence technology in the job market will inevitably lead to a large number of unemployed workers in the short term. Redistributing or retraining these workers will be a major test for governments; the historical experience of the first, second and third industrial revolutions combined shows that the most advanced technology will be in the hands of the few, and the serious imbalance in resource allocation will lead to social conflicts.

In addition, the relationship between governments and monopoly enterprises that have mastery of regulatory technology is no longer equivalent to the relationship between governments and large state-owned enterprises. Government control over the market will be further weakened, and the loss of traditional workers will also lead to a reduction in tax revenue, with adverse effects on the macroeconomy and the regulatory power of governments.

The International Scramble for Jobs

From America's 2016 "Roadmap for US Robotics" and "Brain Research Through Advancing Innovative Neurotechnologies" program, to Europe's SPARC Program and Human Brain Project, to Britain's "Robotics and Artificial Intelligence" report, European and American powers have in recent years released numerous development plans in the field of artificial intelligence.

These reports predict that the large-scale application of smart technology will fundamentally change the employment market in each country, and also affirm that emerging industries will bring new job opportunities to replace the industries that may disappear. At the level of individual workers, they predict that people may change jobs more frequently, requiring them to master skills that can quickly be transferred to different jobs. The documents emphasize the importance of talent and mechanisms for training talent in the future.

In addition, it is not difficult to see from these plans that countries have realized that with the advent of the artificial intelligence era, the flow of labor between countries will become more and more frequent. After all, it is much easier to buy a foreign robot than to hire a foreign employee. Therefore, in the future international job market, whoever has the most advanced technology and the right to formulate international standards will be able to gain greater initiative and flexibility. Therefore, all countries have spared no effort in planning to gain a leading role through promoting the development of artificial intelligence technology.

Disappearing Iron Rice Bowl

There is a view that every technological revolution requires at least one generation to eliminate its negative effects, including the disappearance of industries, the decline in the employed population and the search for a way out for displaced workers. What can be done to minimize this length of time?

International Cooperation and Harmonized Standards

It is foreseeable that after entering the era of artificial intelligence, the gap between different countries' labor markets will be further narrowed, and technical cooperation and data flow between countries will become more frequent. Therefore, establishing uniform technological and testing standards will help increase countries' trust in the international artificial intelligence labor market and facilitate exchange and cooperation between countries. Security standards are especially important, including rules for the collection, processing and cross-border movement of data as well as minimum security standards.

Proactive Government

Faced with the challenges brought by the era of artificial intelligence, the government should separately formulate short- and long-term industrial strategic plans and enact national digital strategies as soon as possible to help workers better cope with increasingly automated and autonomous markets and to prevent the rejection of digitization.

The government should also educate people through publicity efforts and allocate resources according to its plans, increase investment in vocational training, give workers the opportunity to update their skills. In this way it can reduce the negative impact of the large-scale application of automation technology and automated machines on workers' employment and stabilize the job market.

Workers on the Front Foot

From the perspective of every worker, the wave of the times is irreversibly advancing. The tremendous changes in the job market will challenge or even completely overthrow the knowledge structures that this generation developed from childhood, and the new winners will be those who can accept, adapt and lead this transformation most quickly. Therefore, in order not to be left behind by the times, workers need to constantly update their professional skills, keep up with the latest trends in technology development, and learn to engage in emerging industries and occupations in the AI era, such as artificial intelligence designers, engineers and operators. To become first-rate or cross-disciplinary talents who cannot be replaced by AI, honing one's ability to change profession at any time will become the most important criteria of competitiveness.

CHAPTER 30

War Robots

A New Round of Military Revolution and the Birth of Robots

The course of development of human civilization has continuously been accompanied by wars, as small as between tribal factions in primitive societies, and as large as the world wars among dozens of countries in the twentieth century. The haze of war has never disappeared. From the slashing of swords and flesh and blood, to the penetration of artillery and smoke, from the torrent of planes and chariots to the duel of data and information, the pattern of war always follows the pace of human industrial civilization and constantly changes. Behind every change is a major military revolution.

Accompanying the technological achievements of the three industrial revolutions, human warfare successively went through three major revolutions from cold weapons to hot weapons, from hot weapons to mechanization, and from mechanization to informatization. Every major revolution has caused the new capacity to wage war to far exceed the old capacity to wage war, leading to the next new military competition and technological revolution.

The movie *The Last Samurai* tells this story: a dejected American retired military officer is hired by the Meiji government in Japan as a military instructor to train a new-style army to confront the old-style warrior group. After his troops are defeated and he is captured, he gradually integrates with the warriors. At the end of the movie, facing armed US

Tencent Research Institute et al. (eds.), *Artificial Intelligence*,
https://doi.org/10.1007/978-981-15-6548-9_30

military officers, the samurai warriors gallop ahead with swords drawn, ultimately falling under the gunfire of the new Maxim machine gun. This story represents the essence of the first military revolution. The sword ultimately cannot beat the gun, and by the late nineteenth century major countries had completed the military transformation from cold weapons to hot weapons.

My Way is a movie based on World War II that tells the story of North Korean soldiers recruited by Japan, from the Kwangtung Army's confrontation with the Soviet Union to becoming German Nazi soldiers who resisted the Normandy landing. In the beginning of the movie, the Japanese army faced the steel torrent of the Soviet tanks. Without any ability to counter, the Japanese army was almost annihilated. In World War II, the huge advantage demonstrated by mechanized forces enabled all countries to begin the second military revolution from hot weapons to mechanized development, which lasted until the end of the twentieth century.

The story of the film *Courage Under Fire* took place during the Gulf War, which was a computerized war of informatization. From January 17 to February 24, 1991, the US-led multinational force carried out 38 days of air strikes on Iraq and paralyzed Iraq's command and control system. The Iraqi military, once the number one in the Middle East and the fourth in the world, was routed across the board; 29 divisions lost the ability to wage combat. US President Bush announced that the multinational force would stop fighting at 8:00 pm on February 28, ending the Gulf War. The Iraqi military suffered casualties of about 100,000 people and 175,000 Iraqi prisoners and the loss of the vast majority of its tanks, armored vehicles, and aircraft. Only 148 US troops were killed and 458 injured, and in the other countries, 192 were killed and 318 injured. This was a very asymmetrical war. The multinational armed forces used vast information technology to complement the advantages of the navy and the air force, and the Iraqi military, battered to the point they became superstitious in the flood of steel, had no power to fight back. On the other hand, the disproportionate casualty figures are in stark contrast to the hundreds of thousands of US casualties in the Korean and Vietnamese wars. The Gulf War has made all countries realize that only the third military revolution of informatization can achieve enormous superiority in warfare.

The history of the evolution of war proves that "military reform is an endless road that will not come to a halt at this station of informatization. After a short stay, it will pick up a good pace, continue to move forward, and will accelerate forward."

Although informatized warfare, compared to traditional warfare, reflects a tremendous advantage, it still cannot break through a bottleneck—the casualties of combatants. In the Gulf War, the representative example of informatized warfare, more than 1100 casualties of the US-led multinational coalition forces occurred. In the "War on Terror", the most prolonged conflict in US history, according to 2011 data since the beginning of the war on terrorism in 2001, the number of officers and soldiers killed in the battlefields in Afghanistan and Iraq alone has exceeded 6000. There are also tens of thousands of disabled personnel, and this brings a greater number of disability pensions.

Enormous casualty figures not only left politicians bruised and battered, forcing Western countries to change their military strategy and political planning, but they are also spawning the next major military revolution. So, in which direction will the next military revolution after informatized warfare develop? Since the beginning of the twenty-first century, we can see in the war on terrorism a clear outline of the transition from informationization to unmanned and intelligentized transformation. An important symbol of this revolution is the birth of a war robot.

The current form of warfare is on the eve of a major revolution. Military experts have already predicted that an era of information warfare is ending and that another military revolution is about to begin. Military experts at the National Defense University think: "Accompanying the 'rapid vanguard' of the knowledge revolution—the rapid development and application of information technology—there are many new and difficult problems that have arisen in the military field, resulting in a large number of new demands. The battle 'crown' will be exchanging places." Perhaps the war robot is the diamond on that crown.

R&D Trends in Autonomy and Intelligentization

As early as World War II, the German army used mine-sweeping and anti-tank remote controlled explosions, and these became the prototypes of the first war robots. With the rapid development of science and technology, especially since the rapid development of remote sensing, communication, autonomous operation and control technology, and artificial intelligence technology since the 1990s, the development and application of war robots have attracted much attention in many countries. In fact, in the war on terror, war robots have begun to enter the battlefield to assist soldiers and even independently carry out combat missions. War robots are playing an increasingly important role in the battlefield.

The development of war robots from their birth to the present can be broadly divided into three stages: the remote-controlled task phase, the semi-autonomous combat phase and the autonomous unmanned combat phase. In the remote-controlled task phase, a professional operates a remote control device, controlling the robot from a long distance to execute a task. Semi-autonomous robots perform tasks intelligently under the supervision of personnel, but because their intelligence level is not high, they may encounter difficulties in the execution of tasks and require remote intervention from a human to accomplish the expected tasks. Autonomous robots have a high degree of intelligence. The intelligence of the navigation system and the recognition system is sufficient for them to successfully avoid obstacles, recognize both the enemy and their fellow soldiers, and take the initiative to perform their tasks without human intervention.

At present, war robots have not yet achieved full autonomy and require operation by the controlling personnel prior to firing. However, the degree of autonomy of war robots is constantly improving. If this trend continues, the need for human beings to manipulate robots may fade out, and it may even be that the full autonomy of robots will be realized. The chief US Air Force scientist even predicted: "By 2030 machine capabilities will have increased to the point that humans will have become the weakest component in a wide array of systems and processes."

Notable Combat Superiority

The emergence and development of war robots will have a significant impact on the combat methods and characteristics of future battles. In actual combat applications, war robots have significant advantages: first, they have higher intelligence and autonomy; second, all-domain, all-weather combat capabilities; third, strong battlefield survivability; fourth, absolute obedience to orders; fifth, lower operating costs.

In addition, war robots also display more strategic advantages. One is to extend the field of combat space and improve combat effectiveness. As unmanned combat aircraft, unmanned submarines, and space robots have been developed and applied one after another, the scope of combat operations has been extended to high altitude, deep sea, space, and other areas. This makes it possible not only to strike the enemy at long distance, but also to attack strategic targets in enemy territory, beyond the line of sight. In addition, due to the integration of artificial intelligence technology, war robots possess a certain degree of independent combat capability that can withstand the most dangerous and difficult combat tasks which human soldiers cannot bear.

The second advantage is significantly reducing casualties and lowering operational costs. The most notable feature of war robots is that they are unmanned, so command and control staff will be outside the battlefield, controlling the course of combat using remote sensing technology. The introduction of war robots can greatly reduce the casualties of combatants and the wasting of war resources.

The third advantage is enhancing comprehensive combat strength. The military applications of war robots cover almost all areas of operational demand. They have strong battlefield adaptability, with the ability to be used in a variety of operational environments and for various types of warfare. They can operate independently or in synergy for combat. They possess all-weather, all-day, all-domain combat capability. With the intelligentization, synthesization, integration, and standardization of its platform control, the operational concept of the war robot swarm is already being used in actual combat. It is possible to use a cluster of Unmanned Aerial Vehicles (UAVs) to go deep into enemy lines and conduct assault operations in harsh environments. This can defeat opponents with the help of the element of surprise.

The Sword of Damocles

In the past ten years, the ethics of robotics have become a hot topic for discussion. Within this topic, the ethical problems of war robots have led to wide-ranging discussions. Under the historical background of the rapid development of modern computer and artificial intelligence technology, research and development into the automation of war robots has drawn great attention from all countries. Yet, once such a robot is born, it will not only completely change the rules of war, but also challenge the ethical red lines of humanity. Is life and death for a robot to decide? What kind of norms can bind war robots? The development trajectory of war robots has provoked alarm.

The war robots that are currently in combat still have to be controlled by humans, just like remote-controlled toys. They are only machines, whose goals, routes, and actions are determined by humans, especially when it comes to achieving the ultimate function—the use of lethal force. However, it appears this will change soon. Over the past decade, the plans and road maps of all US forces have clearly demonstrated the desire and intention to develop and deploy autonomous war robots. For air, ground, and underwater vehicles, these plans to expel humans from the control system are already underway. And the United States is not the only country focusing on the development of autonomous war robots. Although the border patrol robots developed and used by South Korea and Israel are mainly taking on autonomous monitoring functions, some have pointed out that these types of robots actually have automatic modes and can decide themselves whether to fire. Judging from the current trend of development, the ultimate goal of countries in developing future war robots is to bring about a battle network of war robots covering the ground, sea, and air. Working in synergy, they will fight independently to find targets and destroy them, without the need for human intervention.

As the degree of autonomy of war robots continues to rise, it is plausible that they will be heavily deployed in national armies and replace human beings as the main force in future battlefields. However, compared with human soldiers, war robots raise several ethical issues that cannot be ignored, including the efficiency and obedience of war robots as soldiers and their being accurate and deadly weapons.

One consideration is that war robots have no sympathy and fear. They are tireless; their only goal is to complete the combat mission. They will not show mercy to hostile targets. They are absolute killing machines. As high-tech weapons, the large number of civilian casualties caused by war robots has been much criticized. Take the US UAVs: the number of US military attacks on Afghanistan hit an average of 33 per month in 2012; more than 330 attacks were launched in Pakistan. According to statistics, 35 percent of the victims of drone deaths in 2011 were civilians. Instead of being in the battlefield in the midst of smoke and bloody fights, combatants operating unmanned combat aircraft are throwing bombs remotely across the screen as if playing video games. This minimizes the negative psychological effects of killing, and the restraints of humanity and morality in warfare gradually dim.

This is still the case when human remote control is required, and as war robots become more and more autonomous, it is hard to imagine what humanitarian catastrophes they may cause in warfare once they have the capacity to choose their own goals and fire independently. To this end, some experts in the United States have started to study the ethics to be followed by autonomous war robots. They have tried to implement moral code into existing autonomous robotic systems, that is, to let the robots have an "artificial conscience" in order to achieve precise control over the ethics of wars involving robots.

Although scientists have begun to inject ethical considerations into the development of autonomous war robots, procedures are ultimately mere procedures, and Murphy's law tells us that "Anything that can go wrong will go wrong." Even a more-than-perfect design may produce unexpected situations. So, if a war robot, even an ethically designed autonomous war robot, makes mistakes on the battlefield, who should be punished, and who should bear the responsibility? This is an unavoidable ethical predicament faced by autonomous war robots.

Another consideration is that over history, a hard to ignore reason for why previous human wars have been able to come to an end is the huge losses, especially casualties, caused to both sides. The reason for the end of World War II was precisely that based on US military analysis, sending soldiers to attack Japanese soil might produce a painful toll of 1 million casualties, and so it sought a more effective way to end the war. It ultimately launched two atomic bombs on Japan. Japan, previously prepared to have its soldiers "die in glory" to get the US military to the negotiating table, unconditionally surrendered due to the enormous casualties caused by the atomic bomb. In 1975, the 14-year long war in Vietnam ended with the withdrawal of US troops, following the killing and wounding of 340,000 US troops and a resulting wave of domestic antiwar activities. In 1993, the US military withdrew from Somalia after the "Black Hawk Down" incident in which US helicopters were shot down. In 2009, Obama announced the withdrawal of troops from Iraq, due to the antiwar feeling provoked by the war casualties.

In the future, the large-scale introduction of war robots into the battlefield will make warfare a "breeze" and the "zero casualties" and low-cost advantages of war robots will greatly reduce the voices opposing war and eliminate the shackles restraining politicians. Without any constraints, war may be more likely to happen and last longer. Inevitably, international rules will once again return to the supremacy of hegemony, and human society may once again be trapped in the quagmire of war. After a nuclear weapon, a war robot might become another sword of Damocles hanging over humanity.

Prevent the Mutation of War Robots

As an important war invention, we have already seen the far-reaching impact that war robots are having on modern warfare and the international community. The resulting ethical dilemmas also make war robots increasingly resemble a Pandora's box that humans should not open.

War robots will not make warfare more humane and moral, but will only take warfare further down the road of dehumanization. In order to prevent war robots from actually changing into the "Terminator" of human civilization, it is necessary for the international community to work together now, limit the development and proliferation of war robots, or at least stop the autonomization and intelligentization of war robots. In particular there should be a total ban on research and development of war robots with independent killing functions. "We must continue to ensure that human beings make the moral decisions and maintain direct control of lethal force."

Bibliography

Cheng, Dongfang, Ning Shan, and Jian Zhang. "军用机器人发展趋势". *黑龙江科技信息*. 26 (2014).

Docherty, Bonnie. "Losing Humanity: The Case Against Killer Robots". (2012).

Du, Yanyong. "现代战争机器人的伦理困境". *伦理学研究*. 5 (2014).

Huang, Yuancan. "国内外军用机器人产业发展现状". *机器人技术与应用*. 2 (2009).

Kumagai, Jean. "A Robotic Sentry for Korea's Demilitarized Zone". *IEEE Spectrum*. 44: 33 (2007).

Pang, Hongliang. *智能化战争*. Beijing: National Defense University Press, (2014).

Sharkey, Noel. "Cassandra or the False Prophet of Doom: AI Robots and War". *IEEE Intelligent Systems*. 23: 4 (2008): 14–17.

Sharkey, Noel. "The Evitability of Autonomous Robot Warfare". *International Review of the Red Cross* (2012).

Wang, Wenfeng and Xu Xijun. "即将到来的无人化战争". *未来与发展*. 8 (2011).

Zhou, Qiao. "“杀人机器人”引发人类警觉 美无人机肆意屠杀平民". 13 June 2013. 20 March 2020, http://news.ifeng.com/gundong/detail_2013_06/13/26359581_0.shtml.

http://enjoy.eastday.com/epublish/gb/paper264/18/class026400002/hwz1050728.html.

http://mil.news.sina.com.cn/2011-10-05/1300668208.html.

CHAPTER 31

Soulmate

Scientist Stephen Hawking once said with great enthusiasm: In the future, artificial intelligence can develop a will of its own, a will that will conflict with us. Once artificial intelligence has broken out of its constraints, it will redesign itself in a state of constant acceleration, whereas human beings, limited by the long timeframe of biological evolution, will have no way to compete and will be replaced. We cannot know if we will get the help of artificial intelligence forever, or be despised, marginalized, or even destroyed by it. In short, the success of artificial intelligence may be the biggest event in the history of human civilization, but artificial intelligence may also be what puts an end to that history. From IBM Watson defeating human champions in the quiz show "Jeopardy," to AlphaGo defeating human Go masters, to applications of artificial intelligence in employment and on the battlefield, people cannot help but ask: Will human beings be repressed by artificial intelligence in the future? Will humans and artificial intelligence fall into an endless state of mutual destruction?

We may find the answer from the French film "Her". Ever since the male protagonist broke up with his beautiful girlfriend, he had been unable to escape the shadow of that relationship, and his romantic life had not been smooth. But one day, an artificial intelligence girlfriend who he could only hear but not touch unlocks his long-closed heart. She has a charming voice and is gentle and considerate as well as funny. They soon discover that they have a connection; although they cannot feel the heat and breath of each other's bodies, the communication of their souls brings them long-lost warmth. Although the male protagonist and his virtual girlfriend

Tencent Research Institute et al. (eds.), *Artificial Intelligence*,
https://doi.org/10.1007/978-981-15-6548-9_31

do not live happily ever after at the end of the movie, the idea that artificial intelligence can open its heart to humans has triggered the imagination and longing of countless people. So perhaps we have reached a conclusion that there will be another expectation for the relationship between human beings and artificial intelligence: mutual reliance and love, with artificial intelligence becoming our soulmates and accompanying us around even for our whole lives. Some people have even wondered: Sanmao, Hemingway, Leslie Cheung, and Qiao Renliang all suffered from depression, but if artificial intelligence had become their soulmate, would they have been able to avoid this tragedy?[1]

A Master at Reading Minds

Artificial intelligence is imitating humans in many ways: thought, speech, and movement. The machine is more and more like a living person, and this is why the Turing test exists: can a machine fool people with its own way of thinking? How to make artificial intelligence better understand human emotions is an important part of the Turing test. At present, some laboratories around the world are conducting such research, hoping to develop more smartphones and chatbots that can read human emotions. Letting robots instantly understand people's minds has always been the general direction of artificial intelligence efforts. However, human wisdom is vast. Even if all the knowledge from ancient times to the present were stored in chips, a robot would still need to learn to communicate in order to demonstrate value, and the primary task in communication is to recognize human emotions.

Some progress has been made in this area, such as facial recognition technology. When you play a game, the technology can detect changes in facial expressions and find out whether after reaching a certain level you have become tired. The Japanese emotional robot Pepper is equipped with speech recognition technology, joint technology that enables graceful posture, and emotion recognition technology. It has the most intuitive sensory system that humans can understand: auditory, tactile, and emotional systems. Nearly 200 emotional applications have been launched on Pepper so far. For example, the Pepper diary can take photos during family activities and write diary entries, storing family members' memories like a smart

[1] Sanmao was a Taiwanese author and translator. Leslie Cheung was a Hong Kong singer and actor. Qiao Renliang was a Chinse singer and actor. All three committed suicide.

album. It can guess the mental state of people at a given time, and then chat with you and tell jokes. More than 10,000 Peppers are currently serving families in Japan and Europe. Meanwhile the Han robot developed by Hong Kong company Hanson Robotics can not only understand the user's emotions, but also express emotional feedback with simulated facial expressions. China's Gowild company has also launched a "Gongzi Xiaobai" robot that can act as a life assistant and provide young people with strong social and emotional communication services.

Know You Like "Her"?

Artificial intelligence personal assistants like the one in the movie "Her" are also becoming a reality. Voice assistants such as Apple's Siri, Microsoft's Cortana, and Google Now are appearing, as well as more service-oriented robots. As technology continues to evolve, robots are getting better at reading changes in human emotions and interacting with us more smoothly.

With emotional computing, AI can accurately identify user emotions through semantics, images, and speech. It can understand the user's true intentions and needs through the context of natural dialogue. It not only uses text, voice, and visual communication, but also has individualized memory, providing a one-to-one personalized service that enables users to have emotional trust in and dependence on "emotional robots". The emergence of emotional robots has broken the pre-existing notion of robots as "ice-cold" and will bring warmer, more humanized services to people. In the future emotional robots will slowly permeate family life. Groups such as empty nest elderly people, autistic children, and those suffering from health conditions and seemingly ubiquitous loneliness have stimulated the market for emotional robots. At present, the industry is optimistic about the future market for emotional robots. British scientists even predict that by 2050, humans may be "married" to emotional robots tailored to their needs. Perhaps in the future, people who remain single forever will no longer exist.

With the continuous advancement of technology, artificial intelligence has permeated every aspect of our lives. The idea of machines understanding human minds is no longer just a scene from a film. In the future, perhaps we can create robots that understand human emotions from multiple dimensions, and at that time, they will become our soulmates of the new era.

Bibliography

"1万元带回家 你真的可以和情感机器人谈恋爱了". 18 June 2015. http://www.chinaz.com/news/2015/0618/415358.shtml.
"乔任梁去世:如果有了人工智能能否避免他的悲剧?" 18 September 2016. https://tech.sina.com.cn/it/2016-09-18/doc-ifxvyqvy6644216.shtml.
"一人饮酒醉?你可能需要一个情感机器人!". 3 March 2017. http://www.sohu.com/a/127838434_616238.

CHAPTER 32

New Productive Force

The Economic Revolution Driven by Artificial Intelligence

In the two thousand years before the industrial revolution, the living standards of people around the world did not improve much. The late economic historian Angus Maddison's economic studies of civilizations in different historical periods found that the world's per capita wealth did not improve from around the first year AD to the Industrial Revolution of the eighteenth century. But at the time of the Industrial Revolution, everything changed dramatically. In Marx's *Das Kapital*, productive forces are expressed as the ability of man to transform nature. The two industrial revolutions promoted the rapid development of society's productive forces, which eventually replaced the natural economy with the commodity economy. Handicrafts workshops transitioned into factories with large machine production, achieving a huge leap in productivity. The artificial intelligence industry, after 60 years of ups and downs, burst into brilliance in 2016, ushering in the third wave of technological change. The industry has key technologies such as deep learning at its core and is supported by data and computing power, including cloud computing and biometrics. Artificial intelligence has made breakthroughs in many aspects, and the development trend of artificial intelligence has been in full swing globally. There is no doubt that the era of artificial intelligence has arrived. Just as the steam engine replaced horses as a source of motion, artificial

Tencent Research Institute et al. (eds.), *Artificial Intelligence*,
https://doi.org/10.1007/978-981-15-6548-9_32

intelligence as the new productive force will also bring about earth-shaking changes in various industries and spark new transformations in productivity.

A New Round of "Apollo Missions"

At present, developed countries have deployed artificial intelligence strategies, and hope to use artificial intelligence to promote rapid economic development and create new economic legends.

The human brain is a huge system with extremely complex functional structures. In order to solve gaps in knowledge about the brain that have existed for thousands of years, many countries have proposed "brain plans." In 2013, the European Union proposed the Human Brain Project (HBP), which will last for ten years. The EU and participating countries will provide nearly 1.2 billion euros to make it the most significant human brain research project in the world. When it comes to brain research, even small discoveries and improvements can have enormous economic and social benefits. Through the integration and simulation of data, a better understanding of the structure and functions of the human brain can help lead to innovative treatment programs for brain diseases, such as strengthening diagnosis and treatment of Parkinson's disease, Alzheimer's disease, and so on. This will improve the favorable position of the European pharmaceutical industry in the field of new drugs for brain diseases worldwide. Brain science research is a project with high-tech added value. It is foreseeable that research on brain science will transform the industrial landscape in the future, which would definitely drive the development of related industries and generate huge economic benefits.

The industrial prospects for human brain engineering are very broad, and the prospects for making money are unlimited. On April 2, 2013, US President Barack Obama announced the launch of a program called "Brain Research through Advancing Innovative Neurotechnologies," with $100 million of funding in the first year. The main purpose of the plan is to explore the mechanisms of the human brain and develop the treatment methods for currently incurable brain diseases. Studying the mechanisms of the human brain is not only crucial for treating brain-related diseases, but also holds revolutionary potential for developing computers similar to human brains, greatly advancing the development of artificial intelligence. Among the participants in this program, the US Defense Advanced Research Projects Agency invested about $50 million in 2014 with the goal of understanding the dynamic functions of the brain and using these findings to open up new applications. The agency has established cooperation with technology companies such as Google and IBM and secured a number of important scientific research achievements in artificial intelligence. The US's National Institutes of Health invested about $40 million in developing new technologies for brain research, while the National Science Foundation invested about $20 million to support brain research across disciplines including the physical sciences, biology, and the social and behavioral sciences. After the plan was released, the government, enterprises, and university research institutions attached great importance to it and actively promoted it. Breakthroughs have already been made in many aspects.

In January 2015, the Japanese Ministry of Economy, Trade and Industry released "Japan's Robot Strategy: Vision, Strategy, and Action Plan," with the aim of achieving progress and breakthroughs in its robotics industry. The strategy not only stated a need to increase support for innovation and research and development, but also emphasized the promotion of robots in industry, with the aspiration of becoming the country with the most extensive application of robots. From the formulation of the Japanese strategy and its specific content, it is evident that the Japanese government regards robotics as an important growth area for future economic development, and is striving to promote the Japanese robotics industry internationally.

Artificial Intelligence: New Factor of Production

In today's world, the ability to use capital investment and the workforce to drive economic development has dropped significantly. These two levers are traditional production forces, but in most developed economies they are no longer sufficient to maintain continuous prosperity. However, we do not need to be too pessimistic. In the new stage of development, a new factor of production, artificial intelligence, has begun to step onto the international stage. Artificial intelligence can overcome the limitations of capital and labor, bring new value, and develop resources. Nowadays, the Internet has developed into the Internet of Everything, and the explosive growth of data has triggered the need for effective filtering of information and rational allocation of resources. In an era in which everything is interconnected and computed, productivity growth will accelerate exponentially and drive a new round of industrial innovation. The era of artificial intelligence is centered on key technologies such as deep learning and supported by types of data or computing power such as cloud computing and biometrics. It will see artificial intelligence take root in areas of application such as finance, medicine, autonomous driving, security, the home and marketing, and create huge economic value. Moreover, in the future, artificial intelligence will gradually expand from relatively professional fields to numerous areas of life, and transform into general intelligence to promote a new industrial revolution.

In a nutshell, there are two main reasons why artificial intelligence can promote a new industrial revolution: supercomputing power and the development of big data. Currently, in order to accommodate deep learning's need for extremely large computing power, supercomputers have become a tool for training various deep neural networks. The deep learning tsunami is building a recursive loop of artificial intelligence. As a branch of machine learning originating from artificial neural networks, deep learning is now not just a multilayer perceptron, but is a series of architecture techniques and methods that can be integrated and differentiated. Specifically, through deep learning algorithms, programs use data models to analyze a large amount of data and continuously learn independently, gradually becoming more powerful.

Artificial intelligence can promote productivity as a new factor of production, because conventional production activities can be replaced by automation. Artificial intelligence can help employees to unleash greater abilities and liberate employees to engage in jobs involving higher creativity and value-add. Capital-intensive industries such as manufacturing and transportation are more likely to benefit from the development of artificial intelligence, as multiple jobs in both industries can be replaced with automated operations.

The Economic Dividend from Artificial Intelligence

Artificial intelligence drives economic development mainly through the following three ways: First, it can create a virtual workforce, that is, "intelligent automation." Second, artificial intelligence can improve the skill level of the existing workforce and ensure more efficient use of physical capital. Third, like other technologies, artificial intelligence can promote economic innovation. Over time, it will be a catalyst for a wide range of structural transformations, because artificial intelligence is not only able to use different methods to complete tasks, but can also complete many tasks that did not previously exist.

At the 2017 Summer Davos Forum, PricewaterhouseCoopers and Accenture released reports on artificial intelligence. In "Sizing the Prize," PricewaterhouseCoopers claimed that artificial intelligence will create the biggest business opportunities in today's fast-growing economy. Driven by artificial intelligence, global GDP will increase by 14% by 2030, equivalent to $15.7 trillion. More than 50% of the increase will be attributed to the increase in labor productivity, while the remainder is mainly due to the increase in consumer demand stimulated by artificial intelligence. In terms of geographical distribution, China and North America are expected to be the biggest beneficiaries of artificial intelligence, with a total benefit equivalent to $10.7 trillion, accounting for nearly 70% of the global growth. In 2027, after China has completed a relatively slow accumulation of technology and expertise, it will begin to catch up with the United States. Some developed countries in Europe and Asia will also benefit from artificial intelligence and achieve substantial economic growth. In developing countries, the adoption rate of artificial intelligence technology is expected to be relatively low and corresponding economic growth moderate in scale.

Accenture's report, "How Artificial Intelligence Can Drive China's Growth," suggested that artificial intelligence has the potential to transform China's working methods and open up new sources of value and growth. It could increase the annual growth rate from 6.3% to 7.9%, which translates to an additional gross value added of $7.1 trillion in 2035. The report further analyzed the possible economic impact of artificial intelligence on 15 industries in China. Those that benefit the most from artificial intelligence applications will be: manufacturing; agriculture, forestry, and fisheries; and wholesale and retail. By 2035, artificial intelligence will boost the annual growth rate of these three sectors by 2%, 1.8%, and 1.7% respectively.

Bibliography

Accenture. "How Artificial Intelligence Can Drive China's Growth". 31 May 2019 https://www.accenture.com/cn-en/insight-artificial-intelligence-china.

Lun, Yi. "人工智能各国战略解读:美国推进创新脑神经技术脑研究计划." *电信网技术*. 2 (2017): 47–49.

PricewaterhouseCoopers. "Sizing the Prize: PwC's Global Artificial Intelligence Study: Exploiting the AI Revolution". 31 May 2019. https://www.pwc.com/gx/en/issues/data-and-analytics/publications/artificial-intelligence-study.html.

Yang, Jie and Yao Caifu. "人工智能各国战略解读:欧盟人脑计划". *信息网技术*. 2 (2017): 50–51.

Zhao, Shuyu. "人工智能各国战略解读:日本机器人新战略". *信息网技术*. 2 (2017): 45–47.

Zhou, Chengfang. "论工业革命的社会后果". *内蒙古大学学报*. 1 (1989).

MIX
Papier aus verantwortungsvollen Quellen
Paper from responsible sources
FSC® C105338

If you have any concerns about our products,
you can contact us on
ProductSafety@springernature.com

In case Publisher is established outside the EU,
the EU authorized representative is:
Springer Nature Customer Service Center GmbH
Europaplatz 3, 69115 Heidelberg, Germany

Printed by Libri Plureos GmbH
in Hamburg, Germany